INDUSTRIAL METALS SERIES

Nickel and its Alloys

THE INDUSTRIAL METALS SERIES
General Editor: Dr. E. G. West

Copper and its Alloys
Dr. E. G. West

Nickel and its Alloys
Dr. W. Betteridge

Zinc and its Alloys
S. W. K. Morgan

INDUSTRIAL METALS SERIES

Nickel and its Alloys

W. Betteridge
D.Sc., F.Inst.P., F.I.M.
Consultant, formerly Chief Scientist,
International Nickel Ltd.

MACDONALD AND EVANS

MACDONALD & EVANS LTD.
Estover, Plymouth PL6 7PZ

First published 1977

ISBN 0 7121 0947 1

Printed in Great Britain by
Butler & Tanner Ltd, Frome and London

Introduction to the Series

This series of books covers the metals of industrial importance, other than iron and steel on which there are already numerous texts, giving full details of production, properties and uses. The older non-ferrous metals—copper, lead, zinc, tin—have been neglected by many teachers and therefore by students, with the result that their position and capabilities today are often overlooked. The newer metals, developed within the past century, are often better known but, in general, they have not been the subjects of systematic textbooks.

This series presents up-to-date information on the metals in commercial use. They are intended for students of metallurgy and materials, and those concerned with the engineering disciplines, design and architecture. They can be used not only as textbooks but also as a means of reference for those in industry who require accurate data on the metals which have wide-ranging applications.

Each book, written by an acknowledged expert in the metal concerned, follows the same general pattern. The series has been planned and edited by a former President of the Institution of Metallurgists who, having had wide experience in the application of metals, has been able to direct attention to those aspects known to be of interest to metal users as well as students. Additional books may be added to the series as other metals become of industrial importance and as some of the non-metals become increasingly adopted as alternatives to traditional metals and alloys.

Each book commences with a summary of the properties of the metal—chemical, physical and mechanical—thus providing an immediate means of reference. The subsequent chapters outline the production of the metal from its ores, the family of alloys based on the metal, the processes applied to form and fabricate the metal and its alloys, their metallographic features and their applications, including the economic factors involved.

The readership has been assumed to have some knowledge of the basic sciences on which metallurgy is based, namely chemistry and physics to about the "A" level of the G.C.E. or the level required for H.N.C. and H.N.D. students. A knowledge of the interpretation of equilibrium diagrams has been accepted, but much of the substance of the book can be understood and applied even without such backgrounds. References are included to encourage further reading and study in depth of particular facets.

The purpose of using metals in industry is to provide, at minimum cost, items and assemblies which will fulfil the requirements of the user—be they coins or can openers, automobiles or aeroplanes, buildings or books, toys or tools. Satisfaction in all and every application depends on reliable data about the metals employed. The purpose of the series is therefore to provide a permanent source of information and a reliable guide.

It should be noted that throughout this series the names of alloys which refer to proprietary materials or are trade names commence with a capital letter.

Preface

Among the metals unknown to the ancients but isolated and identified during the technological era, nickel currently stands in a position of great importance in terms of industrial usage—whether measured by tonnage or by market value. In the two centuries or more since its discovery, the scale of production has increased steadily, in spite of fluctuations in demand caused by wars and by industrial recessions, and its regular market price has, in recent years at least, maintained a reasonably uniform level in relation to the general index of prices. Only during short periods of abnormally high demand has the free market price risen significantly above the primary producers' price. This stability is a tribute both to the enlightened marketing policies of the principal mining companies and to the wide range of industries which use the metal. It augurs well for a continuance of the growth of the nickel industry.

This book is intended to outline the whole technology of nickel from mining to final application, at a level which should meet the initial requirements of students, teachers and engineers. It should form a satisfactory basis for the more detailed study of particular aspects of the metal and its alloys, and may also provide, for management and economists, an appreciation of their technological importance.

Much of the information and data included in the book is taken from the publications of Inco Europe Ltd. (formerly International Nickel Ltd.), of Henry Wiggin and Company Ltd., and of their parent company Inco Limited. Grateful thanks are made to them, since without these sources it would not have been possible to compile this book.

1977 W. Betteridge

Contents

List of Illustrations

List of Tables

CHAPTER ONE

Physical and Chemical Properties

In the periodic classification of the elements, nickel occurs in the first long period as one of the three ferromagnetic metals, iron, cobalt and nickel, forming the transition group VIII. The metal appears up to the extent of about 15 per cent in metallic meteorites, and in that form was occasionally used by prehistoric man. Alloys of copper, nickel and zinc resembling present-day nickel-silver were produced in China in the Middle Ages and perhaps much earlier, but it was not until 1751 that Cronstedt working in Sweden isolated and identified the element. He named it nickel, derived from the term *Kupfernickel* (false copper) used by German miners to describe the ore in which it was found. Cronstedt studied the element in a careful and systematic way, laying the foundation for its continued scientific study and the growth of an important metallurgical industry.

Nickel is now one of the major industrial metals. Annual world consumption was about 440 000 tonnes in 1970. At the current market price of about £1430 per tonne (1972) this represents a total value of well over £600m per annum. During the present century the growth of nickel consumption has maintained, over the longer term, an average annual growth-rate of between 6 and 7 per cent, and the industry has geared its exploration, mining and extraction facilities to meet a continuing growth of this nature.

PHYSICAL PROPERTIES

It must be noted that many of the physical properties quoted, and indeed some of the chemical characteristics, are very dependent both on the purity and on the previous metallurgical history of the specimen on which they are determined, these particulars not always being available. It is believed, however, that the data given are reasonably representative of the highest-purity material normally available—usually at least 99·95 per cent. Fuller details of many of the properties can be found in the standard collections of metallurgical data and elsewhere[1–6].

Atomic and nuclear properties

The chemical symbol and atomic properties of nickel are as follows:

Symbol	Ni
Atomic number	28
Atomic weight	58.71

The naturally occurring element is composed of five stable isotopes with the following approximate abundances:

Ni^{58}	67.7%
Ni^{60}	26.2%
Ni^{61}	1.25%
Ni^{62}	3.66%
Ni^{64}	1.16%

Seven artificial isotopes have been identified with half-lives given as follows:

Ni^{54}	0.16 seconds
Ni^{56}	6 days
Ni^{57}	36 hours
Ni^{59}	10^5 years (approx.)
Ni^{63}	85 years
Ni^{65}	2.6 hours
Ni^{66}	56 hours

The reaction cross-sections of nickel and the principal isotopes for absorption of neutrons with a velocity of 2200 m/s are given in Table 1.

TABLE 1

Reaction cross-sections of nickel and the principal isotopes

Absorbent	*Reaction cross-section* (fm^2)
Nickel	460 ± 10
Ni^{58}	420 ± 30
Ni^{60}	270 ± 20
Ni^{61}	180 ± 30
Ni^{62}	1500 ± 300

CHARACTERISTIC EMISSION WAVELENGTHS

The wavelengths of the principal characteristic X-rays emitted by nickel are:

	λ, *nm*
$K\alpha_2$	0.166 17
$K\alpha_1$	0.165 79
$K\beta_1$	0.150 01
$K\beta_2$	0.148 86

The K absorption edge for X-rays is at 0.148 87 nm.

The optical spectrum of nickel is very complex, containing many hundreds of lines, but the sensitive lines of value in qualitative or quantitative spectroscopic analysis are given in Table 2. Over 380 lines are listed by Smithells[2] within the range 989.8–201.4 nm.

TABLE 2

Sensitive lines of the optical spectrum of nickel

Wavelength	*Relative intensity*	
(*nm*)	*arc*	*spark*
352.4541	1000 R	100
351.5054	1000 R	50
349.2956	1000 R	100
341.4765	1000 R	50
228.7084	100	500
227.0213	100	400
226.4457	150	400
225.386	100	300

R = wide self-reversal.

CRYSTAL STRUCTURE

The normal structure of nickel throughout the whole range of temperature up to the melting point is face-centred-cubic, although unstable hexagonal and body-centred-cubic forms have been reported in special cases, e.g. in thin evaporated films or in certain electrodeposits—these are probably due to the presence of impurities. The lattice constant of the face-centred-cubic form is $0.352\,38 \pm 0.0003$ nm at 25°C. The lattice constant at other temperatures can, of course, be calculated using the thermal expansion coefficients given on p. 6.

Density

The theoretical density of nickel calculated from the crystal structure and atomic properties is 8908 kg/m³ at 20°C, but the most reliable practical determinations give values within the range 8900–8908, depending on the purity and the prior treatment of the metal. The density falls with increasing temperature due to thermal expansion and reaches about 8300 kg/m³ at the melting point. The density of liquid nickel at the melting point has been reported as 7900 kg/m³, and with increasing temperature falls approximately linearly to 7000 kg/m³ at 2500°C.

Thermal properties

The melting and boiling points of nickel are 1453 ± 1°C and about 2800°C respectively. The boiling point is imperfectly known since it is normally extrapolated from measurements of vapour pressure at lower temperatures, and these latter are not fully reliable. Figure 1

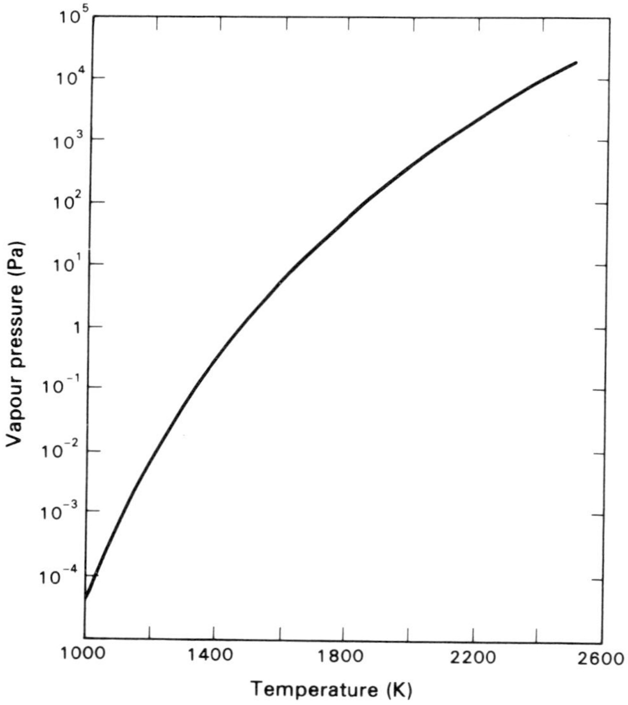

Fig. 1—Vapour pressure of nickel (1 Pa = 0.0075 mm Hg).

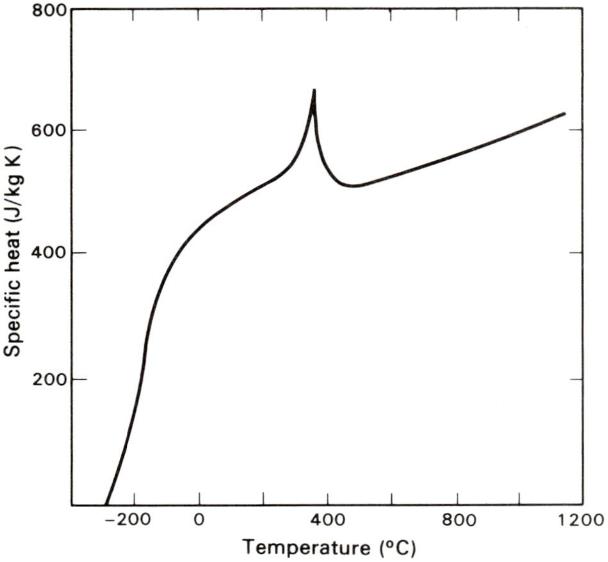

Fig. 2—Specific heat of nickel.

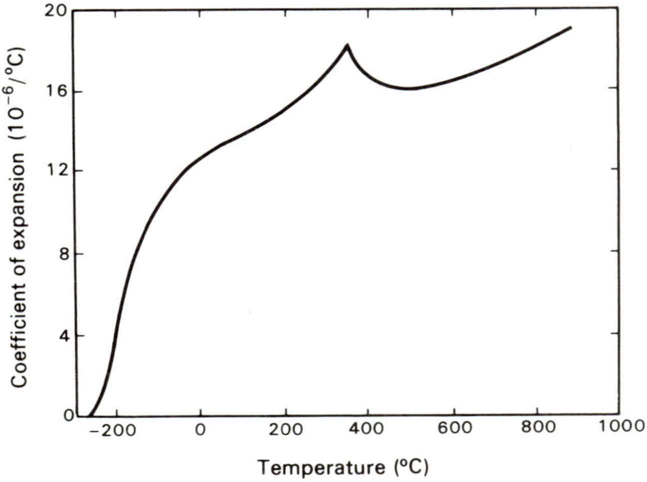

Fig. 3—Thermal expansion coefficient of nickel.

TABLE 3

Coefficient of expansion of nickel.

Temperature range (°*C*)	*Coefficient* (K^{-1})
20–100	13.3×10^{-6}
20–200	13.9
20–300	14.4
20–500	15.2
20–700	15.7
20–900	16.3

shows the results of vapour pressure determinations reported by Honig[7] over the temperature range 1000–2600 K. It is surprising that no inflection is noticed at the melting point, but nevertheless the results indicate the rapid rise in vapour pressure with increasing temperature.

The specific heat of nickel varies within the range 420–620 J/kg K over the range of temperatures between 0 and 1000°C, with a clearly marked peak at or near to the Curie point, the temperature at which nickel loses its ferromagnetism (*see* Fig. 2).

The thermal expansion characteristics of nickel also show, as do many of its physical properties, an inflection at the Curie point (*see* Fig. 3), in which the incremental expansion coefficient is plotted against temperature. The average coefficient over different temperature ranges is listed in Table 3. The thermal conductivity of nickel varies with temperature and again inflects in the region of the Curie point; values at temperatures up to 500°C are given in Table 4.

TABLE 4

Thermal conductivity of nickel.

Temperature (°*C*)	*Conductivity* (*W/mK*)	(*cal/cm s* °*C*)
25	92	0.22
100	83	0.198
300	64	0.152
500	62	0.148

ELECTRICAL PROPERTIES

The electrical resistivity of a metal is one of the most sensitive indicators of purity. It is probable that most of the differences in the quoted values for the resistivity of nickel are due to varying purity of the material tested. Values for pure nickel derived by Pallister[8] from measurements on samples of different purity are given in Table 5. It is to be noticed that the temperature coefficient of resistivity rises to a sharp peak in the vicinity of the Curie point (*see* Fig. 4). Values of resistivity at 20°C of various

TABLE 5

Electrical resistivity of pure nickel

Temperature, T (°*C*)	(*K*)	*Resistivity*, ρ (Ωm)	$d\rho/d$T
−273	0	$0{\cdot}03 \times 10^{-8}$	0.0005
−253	20	0.04	0.001
−223	50	0.18	0.008
−173	100	1.02	0.023
− 73	200	3.70	0.031
0	273	6.15	0.0365
27	300	7.16	0.0385
127	400	11.55	0.0505
227	500	17.5	0.068
327	600	25.2	0.0905
527	800	35.0	0.031
727	1000	40.75	0.027
927	1200	45.85	0.0245
1127	1400	50.6	0.023
1327	1600	55.05	0.022

samples of nickel, ranging in purity from 99.99 per cent to about 99.8 per cent, have varied from 6.8 to 9.9 microhm-cm ($10^{-8}\,\Omega$ m).

While nickel-base alloys are widely used as elements in thermocouples (*see* Chapter 3), the thermoelectric power of the pure metal is not of great practical significance, because oxidation would preclude its use for other than moderate temperatures. Nickel is thermoelectrically negative to platinum, the common reference material, i.e. the current at the cold junction flows from platinum to nickel, unlike many other common metals, such as iron, copper, silver and gold which are positive to platinum. Hence relatively

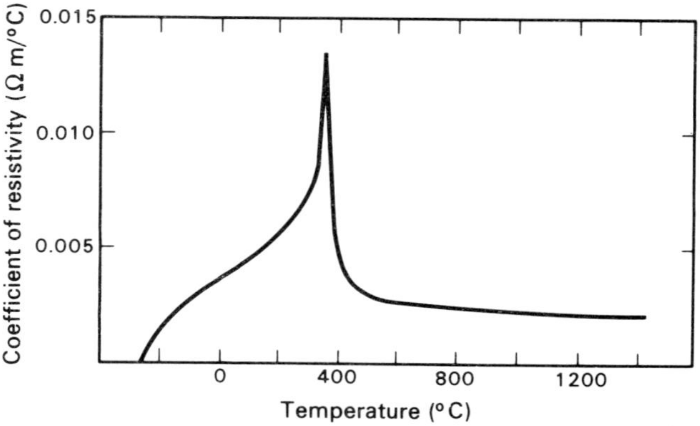

Fig. 4—Temperature coefficient of electrical resistivity of nickel.

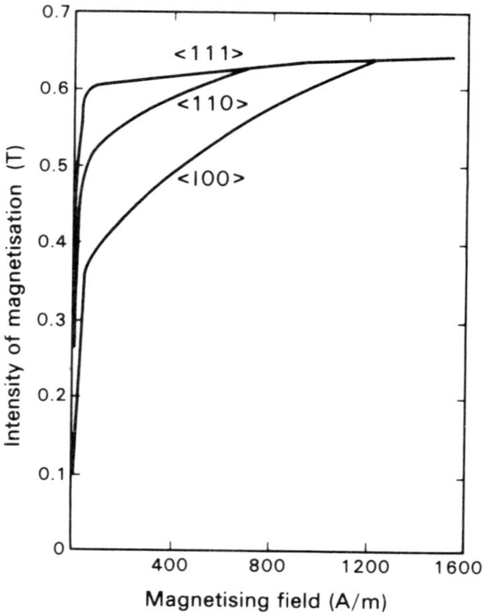

Fig. 5—Magnetisation curves of nickel in different crystallographic directions.

high thermal e.m.f.s can be generated by using nickel as one element, and one of the latter group of metals as the other.

Magnetic properties

Although pure nickel is used because of its magnetic characteristics only in certain special cases, such as for magnetostrictive devices, nickel-containing alloys of many types exploit the particular ferromagnetic properties of nickel. Nickel is an essential constituent of many magnetic materials ranging from high-permeability, soft magnetic alloys to high-coercivity permanent magnet alloys. Although these practical alloys are generally iron-based, they frequently contain 10–20 per cent of nickel or more, and are referred to in more detail in Chapter 3.

The magnetic induction curves for single crystals of nickel show that the direction of easy magnetisation is the ⟨111⟩ direction. the diagonal of the face-centred-cubic unit cell. Figure 5 shows the induction curves for the three principal crystallographic directions. For polycrystalline nickel with no preferred orientation of the crystal lattice, the magnetisation curve at normal temperature is shown in Fig. 6. The saturation induction is 0.617 T (6170 G).

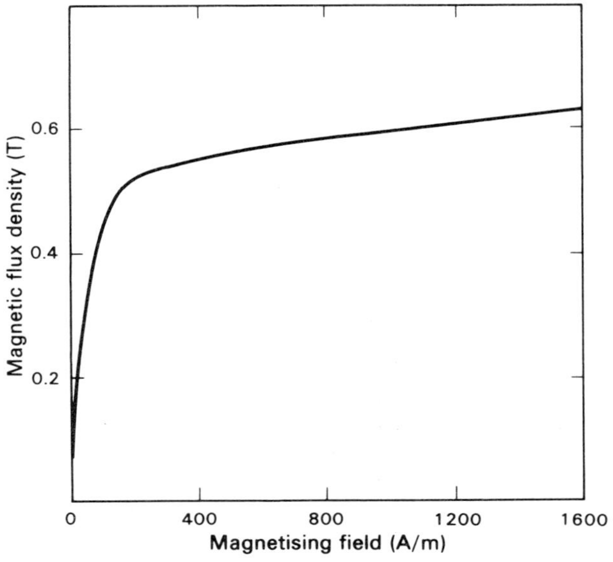

Fig. 6—Magnetisation curve for annealed polycrystalline nickel.

With increasing temperature the saturation induction falls as shown in Fig. 7, until at the Curie temperature, ferromagnetic characteristics are lost and nickel becomes paramagnetic. The Curie temperature, in the same manner as other structure-sensitive properties, is dependent on purity and on the prior history of the specimen, but, in general, for high-purity nickel, lies between 350 and 360°C. For 99.95 per cent nickel, 353°C (626 K) has been quoted.

The initial permeability of pure nickel is about 250×10^{-6} H/m (200 G/Oe) and the maximum permeability of $2500–3500 \times 10^{-6}$

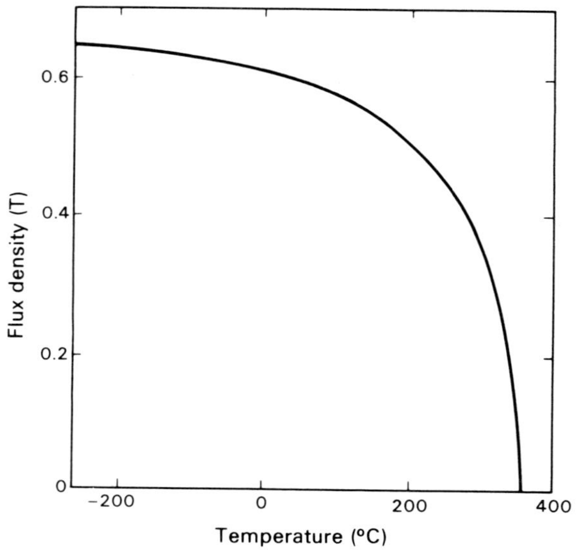

Fig. 7—Saturation magnetic flux density of nickel.

H/m (2000–3000 G/Oe) is reached in a field of about 80 A/m (1 Oe). The coercivity, again greatly dependent on purity and on thermal history, lies in the range 80–250 A/m (1–3 Oe).

The magnetostrictive properties of nickel are of particular practical importance since they are some of the greatest effects shown by a readily available material. The effects are measured by the change in length of a material when magnetised, and the fractional change in length, $\Delta l/l$, is normally only a few parts per million. Whereas iron first expands, and then at higher field-strengths contracts, with a maximum $\Delta l/l$ of about 5×10^{-6}, nickel always contracts when magnetised, the magnetostrictive ratio

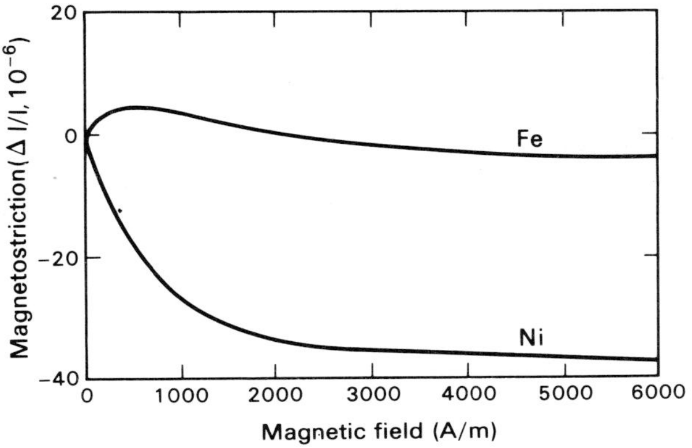

Fig. 8—Magnetostriction of iron and nickel.

reaching almost 40×10^{-6} at saturation. The magnetostrictive behaviour of iron and nickel is compared in Fig. 8. Nickel and nickel-base alloys developed for special magnetostrictive applications are thus widely used as the essential material in a variety of devices such as echo-sounders, ultrasonic generators, strain gauges, etc. (*see* Chapter 7).

Diffusion

The self-diffusion characteristics of nickel, measured by the diffusion of a radioactive isotope in the inactive metal, depend on the grain size and surface effects, since diffusion rates at grain boundaries and surfaces are much higher than those relating to diffusion through the normal unstrained crystal lattice. The lattice self-diffusion coefficient, for the temperature range 850–1100°C, is given by the expression:

$$D = 0.48 \exp(-Q/RT) \text{ cm}^2\text{s}$$

where the activation energy, Q, is 276 kJ/mol (65 800 cal/mol).

Optical properties

The reflectance of nickel at normal temperature, calculated from the fundamental optical constants, increases with increase of wavelength through and beyond the visible range, as shown below:

λ, μm	0.420	0.589	0.75	1.00	2.25
Reflectance	53.8	62.1	70.0	74.2	85.5

The spectral emissivity for light of wavelength 0.65 μm, that relevant to optical pyrometry, is 0.36 for solid nickel and 0.37 for liquid nickel with the metal in an unoxidised condition. The total emissivity, again for unoxidised metal, varies with temperature in the following manner:

Temperature, °C	25	100	500	1000
Emissivity	0.045	0.06	0.12	0.19

CHEMICAL PROPERTIES

The chemical characteristics of metallic nickel which are of principal interest in connection with its practical applications are its good resistance to atmospheric tarnishing and its outstanding resistance to attack by caustic alkalis. The resistance to atmospheric tarnishing was responsible for the metal's first major industrial application as an electrodeposited coating on other metals, particularly steel. Nickel is still very widely used in this manner, both on its own and as an undercoating for chromium plate.

The rate of attack on nickel by solutions or by gases is initially dependent on the surface finish of the metal, because the true surface area can be appreciably greater than the apparent, or projected, area; because of the effects of cold working in modifying the reactivity of the metal; and because of adsorbed films of contaminants. In general, however, the effects of surface disappear as the corrosion proceeds.

Atmosphere

The rate and nature of tarnishing of nickel when exposed to the atmosphere are very dependent on local conditions, particularly the extent of industrial pollution. Although the rate of corrosion is very low, the polished metal slowly becomes dull and develops an adherent film of corrosion product—normally a sulphate—when exposed to outdoor conditions. Indoors, the rate of tarnishing is very slow; for example, nickel-plated parts remain bright, only occasional, light polishing being required.

Results of external atmospheric corrosion-testing of commercially pure (over 99 per cent) nickel extending over 20 years has shown that the attack proceeds uniformly without producing localised pitting, and a constant (linear) corrosion rate is developed in any particular location. The rate of attack varies considerably, being very low in rural and marine locations, but as much as a hundred times greater in industrial locations. The most severe attack observed gave an average penetration rate over a 20-year period of test of only about 5 μm/year.

WATER

Nickel is extremely resistant to attack by distilled water or natural fresh water. In sea-water the rate of attack is low, i.e. less than 25 μm/year, when the velocity of the water over the metal surface exceeds about 1 m/s. Under stagnant conditions, however, sea-water can lead to localised pitting even though the rate of general corrosion is still low.

MINERAL ACIDS

Sulphuric acid. The rate of corrosion of nickel by sulphuric acid depends on the degree of aeration as well as on temperature and concentration of the acid. At normal temperature, the rate of attack is about 50 μm/year for acid up to 10 weight per cent when unaerated, but can rise to several times this figure when fully saturated with air. In boiling acid the rate of attack increases with concentration from 860 μm/year with 5 per cent acid to a maximum of 86 000 μm/year at 50 per cent and falling to 58 000 μm/year at 96 per cent. It must be noted that even the maximum rate quoted still corresponds to very slow attack of the metal, i.e. about 0.01 mm/h.

Hydrochloric acid. Nickel is attacked only very slowly by hydrochloric acid, and again the rate of attack is increased by aeration. At concentrations below 10 weight per cent, the rate of attack at normal temperatures is below 200 μm/year when air-free, but increases to 2000 μm/year when fully aerated. With concentrated acid (25–30 per cent) the rate rises to 1500–2000 μm/year and is much less affected by aeration.

Hydrofluoric acid. The attack of nickel by hydrofluoric acid has been less fully studied, but the pattern of behaviour is similar to that for hydrochloric acid. With acid below about 50 weight per cent, attack in the absence of air is slow, but increases to about 500 μm/year when air is present with 50 per cent acid. Even the worst conditions (35 per cent acid, aerated, 110°C) yielded only a corrosion rate of 28000 μm/year. Anhydrous hydrofluoric acid has very little effect on pure nickel, even at elevated temperatures up to 150°C in a closed bomb.

Phosphoric acid. Phosphoric acid attacks nickel only slowly, and the rate is more controlled by temperature than by acid concentration.

In a solution containing 84 per cent by weight of P_2O_5, corrosion rates were 330 and 1220 μm/year at 60 and 120°C respectively At still higher temperatures, however, excessive corrosion rates were reported.

Nitric acid. Nickel is rapidly attacked by oxidising acids such as nitric acid. It dissolves readily in dilute nitric acid, giving a solution of the nitrate, and it has been reported that in concentrations up to 3.5 N the rate of solution is proportional to the square root of the concentration. In concentrated acid, however, nickel can be passivated.

Alkalis

Nickel is strongly resistant to attack by caustic alkalis, either in aqueous solution or in the molten state; for example, the rate of attack in boiling 50 per cent caustic soda is less than 25 μm/year. This immunity has led to many important applications, such as evaporating vessels in the production of alkalis, and for electrode materials in alkaline storage-batteries. Ammonium hydroxide, however, in concentrations greater than about 1 per cent, does attack nickel, although anhydrous ammonia has no effect. Molten carbonates also attack nickel.

Salts

Attack of nickel by salt solutions is generally restricted to oxidising salts, such as ferric chloride or hypochlorites, and, with these, attack may be severe, but with non-oxidising salts the rates of attack are usually less than 100 μm/year.

Organic materials

In general, nickel may be said to be inert to almost all organic materials. It has, therefore, been widely used as a container material for the processing of many organic materials, particularly foodstuffs, although austenitic stainless steels are increasingly supplanting pure nickel or nickel–copper alloys in this field.

Gases

The reaction of gases with nickel is conditioned by two main factors (apart from the specific nature of the gas)—the temperature and the water content of the gas. At normal temperatures, only wet gases have any appreciable effect on nickel, and those giving acid solutions are particularly liable to lead to attack—oxides of nitrogen, halogens and sulphur dioxide are all in this class. The reactions may then be considered as similar in nature to those outlined for the aqueous solutions of the acids.

Oxygen. Of the reactions between gases and nickel at elevated temperatures, oxidation is by far the most important. When nickel is heated in air or in oxygen it forms an adherent film or scale of nickel oxide, NiO, which controls the rate of further oxidation by acting as a diffusion barrier between the metal and the atmosphere. The mechanism and the rate-controlling process for the initial formation of the film and its growth at relatively low temperatures (say, below 400°C), are not at all clearly understood, but at higher temperatures the growth follows a parabolic law and hence, according to Wagner's theory of oxidation[9], the reaction is controlled by the diffusion of ions and electrons through the protective oxide layer. Thus the thickness, x, of the scale at time t is given by

$$x^2 = at$$

where a is constant at a fixed temperature.

The oxidation kinetics of nickel have been measured most comprehensively by Gulbranson and Andrew[10] and their results are illustrated in Fig. 9. Many other workers have studied the reaction, and a variety of factors influencing the rate-constants have been described. As would be expected from Wagner's theory, the purity of the metal can have a marked effect since the presence in the scale of minor contents of elements with a different valency

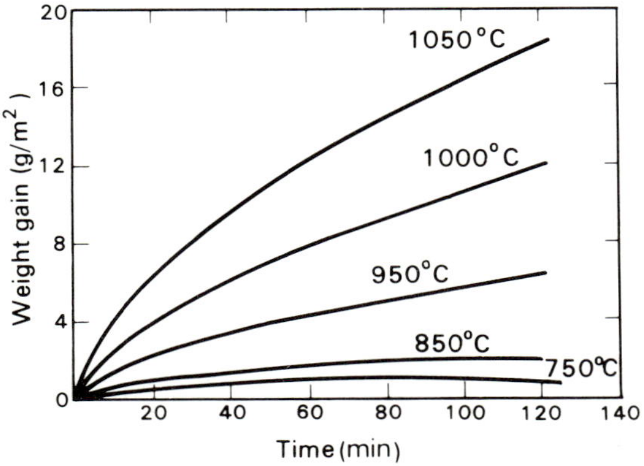

Fig. 9—Oxidation of nickel in pure oxygen at a pressure of 10^4 Pa (76 mm Hg).

to that of nickel can affect ionic diffusion rates. Thus a small addition of lithium (valency 1) increases the oxidation rate at 1200°C by a factor of seven, while small amounts of chromium (valency 3) also increase the oxidation rate, contents exceeding about 7 per cent being required before improved resistance to oxidation is obtained. According to Kubaschewski and Hopkins[11], the parabolic rate-constant, *a*, for the oxidation of nickel is given by the expression

$$a = 3.2\,e^{-\frac{45\,000}{RT}}\ \mathrm{kg^2\,m^{-4}s^{-1}}$$

over the temperature range 750–1240°C. The oxidation rate-constants for pure nickel according to Yao[12] are given in Table 6.

TABLE 6

Oxidation of nickel
((weight change)$^2 = a$.time).

Temperature (°C)	*Parabolic rate constant, a* ($kg^2m^{-4}s^{-1}$)
600	4.22×10^{-11}
700	4.22×10^{-10}
800	5.62×10^{-9}
900	1.78×10^{-8}
1000	1.78×10^{-7}

Sulphur-containing gases. Serious attack of nickel is found with sulphur-containing gases, particularly at a temperature above 643°C, the melting point of a eutectic formed between nickel and the sulphide, Ni_3S_2. Below this temperature, the attack proceeds by a diffusion process forming a surface layer of sulphide which is rate-controlling, but at higher temperatures the liquid eutectic leads to intergranular penetration and serious embrittlement of the metal. The attack is most severe with reducing gases such as hydrogen sulphide, since oxidising gases like sulphur dioxide can also form an oxide layer which is more protective than sulphide layers. Corrosive attack of the same basic nature can arise from the presence of any sulphur-containing compounds during the heating of nickel and many nickel alloys, and precautions to exclude such contaminants are generally necessary.

Carbon monoxide and hydrocarbon gases. The reaction of other gases with nickel usually leads only to solution or adsorption of the gas without the formation of chemical compounds with the nickel. Hydrocarbons or carbon monoxide can react at elevated temperatures, with the solution of carbon into the nickel and with the formation of nickel carbide, although the extent of the reaction is not at all comparable with the corresponding reaction with iron. At moderate temperatures, say 50–150°C, carbon monoxide attacks nickel with the formation of the volatile carbonyl $Ni(CO)_4$, this reaction forming the basis of the carbonyl refining process (*see* Chapter 2).

Nitrogen. Nitrogen and ammonia are inert to nickel and no nitrides of nickel are known.

Halogens. Nickel is reported to be resistant to dry chlorine at temperatures up to 500°C, the reaction rate increasing with increasing temperature, reaching a penetration rate of 15 mm/year at 650°C. The observations imply the formation of a protective chloride film. The attack of fluorine has been studied in short-term tests; dense adherent films of nickel fluoride are formed which are protective at temperatures at least up to 540°C. On the basis of 120-hour tests, it appears that a corrosion rate in the order of 0.5 mm/hour at 540°C would apply.

Hydrogen. Hydrogen readily diffuses through nickel and is dissolved to an extent depending on temperature and pressure. The solubility at one atmosphere pressure is about two parts per million by weight at 300°C, rising to about 8 p.p.m. at 900°C. In liquid nickel the solubility rises from about 37 p.p.m. at the melting point to 44 p.p.m. at 1720°C. These latter figures correspond to volume of gas, when reduced to normal temperature and pressure, almost four times the volume of the metal. The diffusion rate of hydrogen in nickel increases rapidly with increasing temperature, but is at a reasonably fast rate even at normal temperature. The diffusion rate has been reported as

$$D = D_o \exp(-b/T)$$

where $D_o = 1.07 \times 10^2\ m^{-2}\,s^{-1}$ and $b = 5100$ K, corresponding to an activation energy of 42 kJ/mol (10 100 cal/mol).

Electrochemical properties

The solution potential of nickel in a 1 N solution of nickel ions (sulphate) is −0.227 V at 20°C and −0.231 V at 25°C. Passivation of the metal, i.e. change of the solution potential on passage of current to the electrolyte, is markedly affected by minor levels of other elements, particularly sulphur, and use is made of this in the nickel anodes used for electrodeposition of nickel to ensure uniform solution of the metal.

Catalytic properties

Finely divided nickel is an important industrial catalyst, particularly for use in the food industry in connection with the hydrogenation of fats and oils. The catalytic form is usually produced by the Raney process[13], in which a nickel–aluminium alloy is treated with sodium hydroxide to extract the aluminium and leave a fine sludge of active nickel. An essential requirement for a hetrogeneous catalyst of this type is a high ratio of surface area to volume. The catalytic action of nickel in hydrogenation reactions is probably associated with the adsorption of hydrogen onto the active surface, although nickel can usefully catalyse reactions other than those involving hydrogenation.

References

1. Rosenberg, S. J., *Nickel and its Alloys*, National Bureau of Standards Monograph 106, U.S. Dept. of Commerce, Washington, 1968.
2. Smithells, C. J., *Metals Reference Book*, 4th edn., Butterworth, London, 1967.
3. *Metals Handbook*, 8th edn., American Society for Metals, Ohio, 1964.
4. Weast, R. C., *Handbook of Chemistry and Physics*, 50th edn., Chemical Rubber Publishing Co., Cleveland, Ohio, 1969–70.
5. *Gmelins Handbuch der Anorganischen Chemie*, 8th edn., Verlag Chemie, Weinheim, 1961.
6. *Landolt-Bornstein Zahlenwerte und Funktionen*, 6th edn., Springer, 1964.
7. Honig, R. E., "Vapour Pressure Data for the More Common Elements", *Radio Corporation of America Review*, vol. 18, p. 195, 1957.
8. Pallister, P. R., "Resistivity of Nickel", *Metallurgia*, vol. 71, p. 165, 1965.
9. Wagner, C., *Zeitschrift für physikalische chemie*, vol. (B) 21, p. 25, 1933.
10. Gulbranson, E. A., and Andrew, K. F., "High-temperature Oxidation of High-purity Nickel between 750 and 1050°C", *Journal of the Electrochemical Society*, vol. 104, p. 451, 1957.
11. Kubaschewski, O., and Hopkins, B. E., *Oxidation of Metals and Alloys*, Butterworth, London, 1967.
12. Yao, Y. L., *Journal of Physical Chemistry*, vol. 33, p. 741, 1960.
13. Adkins, H., and Billica, H. R., "Raney Process", *Journal of the American Chemical Society*, vol. 70, p. 695, 1948.

CHAPTER TWO

Occurrence and Extraction of Nickel

OCCURRENCE AND NATURE OF ORES

Although in the overall composition of the earth nickel is believed to be a most abundant element, making up about 3 per cent of the whole and being exceeded only by iron, oxygen, silicon and magnesium, it is much less prominent in the earth's crust—less than 0.01 per cent of the whole—and it ranks 24th in order of abundance. Fortunately natural processes have served to concentrate nickel in a few locations to a level at which it is rich enough to mine. Two distinct processes have yielded different types of nickel ore. The first is believed to have taken place at some depth in the earth's mantle by the reaction of sulphur with nickel-containing rocks to form complex sulphides. These sulphides were then transferred by earth movements as the mantle cooled and shrank, and were deposited in the positions from which they are now mined. The second process is one of weathering of exposed surface rocks which leads to nickel being taken into solution and redeposited elsewhere as complex oxides, hydroxides or silicates—ores of this type are termed "lateritic".

Sulphide ores, which at present form the major source of the world's nickel, are generally found in areas where glacial action has removed much of the overburden of weathered rock. The most important ore body is at Sudbury, Ontario, in the Canadian Shield, which was discovered during the laying of the Canadian Pacific Railway in 1883. Here the deposit is spread around the rim of an oval-shaped bowl about 60 km by 27 km, and it is recovered in some areas by open-cast mining and in others by deep-pit mining—even to a depth of 2000 m. The ore from Sudbury contains approximately equal quantities of nickel and copper sulphides, the contents of each of these metals being in the region of one per cent. Also present in the ore is much iron, some cobalt

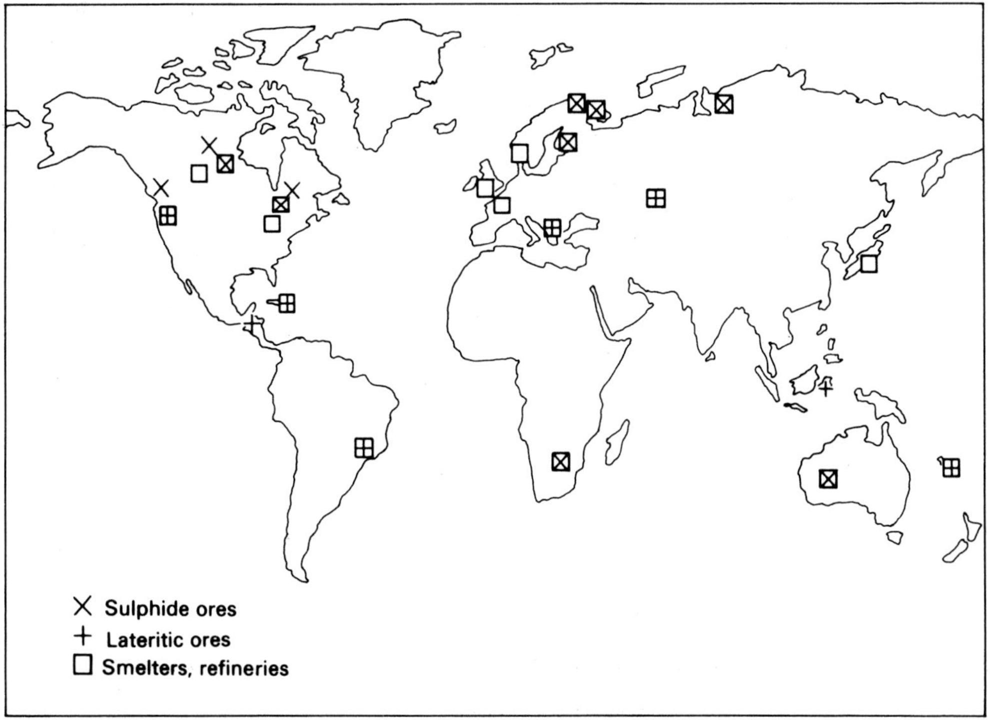

Fig. 10—World location of major nickel mines, smelters and refineries.

and minor but economically important traces of the metals of the platinum group. In all, fifteen elements are recovered from this type of ore. Sulphide ores of essentially the same type, although often with a much lower content of copper, are found in other areas of Canada, such as Northern Manitoba; in the U.S.S.R. at Norilsk in Siberia and at Pechenga near the Finnish border; in Western Australia; in South Africa, where they are mined primarily for the content of the platinum group metals; and also in minor quantities in many other locations.

Lateritic ores, in contrast to sulphides, mainly occur in tropical areas, and vary considerably in nickel content and nature. Depending on the composition, particularly the iron content, the colour of the ore varies from green, through yellow, to chocolate brown. The deposits are generally found within 10 or 15 m of the surface of the ground at elevations between sea-level and perhaps 1000 m above sea-level. The ore is often a soft clay-like material, unlike the hard rock of the sulphide deposits, and hence is recovered by what are essentially earth-moving techniques using power shovels and drag-lines. The nickel content of the lateritic ores originally worked in New Caledonia averaged 9 per cent, but has gradually fallen as the richest ore became exhausted, the average content now being below 3 per cent.

The silicate deposits of New Caledonia were discovered in 1865 and formed the world's principal source of nickel until the Sudbury deposits of sulphide ores became predominant in the early years of this century. They are still a very important source of nickel and indeed increased attention has been paid to them in recent years. Lateritic ores are also mined in considerable quantities in Cuba, the U.S.S.R. and in the U.S.A. (Oregon). Known deposits in Indonesia, Guatemala and elsewhere are being brought to the stage of exploitation. Figure 10 shows the world locations of the principal nickel mines, smelters and refineries.

The proved reserves of nickel are extensive, but they are mainly of lower-grade ores which will be more expensive to exploit than those at present being mined. (For a full account of the geology, mining and extraction of nickel, *see The Winning of Nickel*, Boldt and Queneau[1].)

Lateritic ores are normally worked only for their nickel content which is recovered as metallic nickel, ferro-nickel or nickel oxide. The contents of other metals are not usually of economic interest, although cobalt and chromium may be present in quantities which are likely to be attractive at some future date.

EXTRACTION AND REFINING OF NICKEL

The initial treatment of sulphide and lateritic ores differs completely because of their entirely different physical characteristics.

Sulphide ores

These consist of a physical mixture of the valuable sulphides with a large proportion of worthless rock called "gangue". In order to separate these, the ore is first crushed and then milled as a water slurry until the particles are in the order of 0.25 mm across. To separate the sulphides from the gangue and from each other, two processes may be employed. The first is froth flotation, which depends on the different surface activities of the constituents; the second is magnetic separation, based on the different magnetic characteristics of the materials. Froth flotation is carried out in cells. The slurry of milled ore is passed through the cells while air is blown into it to form a rising cloud of bubbles. Frothing agents and selective surface-active reagents are added so that the required constituents of the slurry adhere to the surfaces of the bubbles and are floated away, while the rejected constituents remain submerged. Successive flotation cells, using different surface-active additions, enable nickel concentrates and copper concentrates to be separated from the unwanted gangue.

Magnetic separators consist of rotating drums carrying the feed material, which may be wet or dry, past stationary electromagnets. The magnetic fractions adhere to the drum surface and are carried away from the non-magnetic fraction. The process is used either as a supplement or an alternative to the flotation process.

The nickel-rich product from these beneficiation processes contains 5–10 per cent nickel after drying.

Pyrometallurgical processes. Most sulphide ores are now treated by pyrometallurgical processes in which the next stages consist of roasting, smelting and converting, which serve to eliminate a great deal of the iron and sulphur, and to increase in steps the nickel content. Roasting is carried out in multi-hearth furnaces, fluidised-bed roasters or rotary kilns, and involves heating the concentrates in a current of air to oxidise much of the iron and remove the bulk of the sulphur as sulphur dioxide—normally without sintering the residue. Smelting raises the material above its melting point and the iron oxide is taken into a siliceous slag which separates from the denser molten nickel sulphide. The operation may be carried out in the older blast-furnaces or reverberatory furnaces, or by the newer processes of flash smelting or arc-furnace smelting. The converting stage carries the process still further, blowing the molten matte with air or oxygen-enriched air to remove virtually all the remaining iron, and yield a matte containing about 75 per cent nickel, mainly as sulphide.

There are now alternative procedures for treating the matte, depending on its copper content. Crushing, grinding and magnetic separation or froth flotation may be necessary to separate nickel and copper sulphides derived from copper-rich ores. The molten matte, if low in copper, may be cast directly into sulphide anodes for electrolytic refining. Alternatively, the nickel–copper sulphide concentrate may be roasted, usually in fluid-bed roasters, to oxide suitable for direct use in steel-making, or partially reduced with hydrogen to give a higher-nickel-content sintered product, also used in steel-making. The oxide may also be fully reduced to crude nickel and used as the feed-stock to refineries for conversion into what is essentially the pure metal. Two main processes for this are used, electrorefining and vapo-refining.

Electrolytic refining. The electrolytic refining process uses anodes cast from the crude metal and a sulphate–chloride electrolyte with an addition of boric acid, which operates at 60°C and a current density of between 1.5 and 2 A/dm^2. Cathode starting sheets are made by deposition on to stainless-steel blanks from which they are stripped after two days, and then built up during a period of about 10 days into plates about 1 cm thick, forming the final basic product with a purity exceeding 99.9 per cent. The electrolyte is pumped continuously through the electrolysing tanks and a purification train to remove soluble impurities, while the anodes are enclosed in fabric bags to retain the insoluble residues, which are subsequently treated for the recovery of their useful constituents.

The sulphide anodes referred to earlier are treated by an essentially similar process. Electrolytic processes are also used for the recovery of nickel from the leach liquors, using insoluble anodes such as antimonial lead.

Vapo-refining. The vapo-refining process is based on the reaction of carbon monoxide with nickel at temperatures in the region of 100°C to produce the volatile carbonyl, $Ni(CO)_4$, which on further heating to about 200–300°C decomposes to give the metal and carbon monoxide. While iron and cobalt also form volatile carbonyls, the conditions of temperature and pressure used in nickel refining are such that the formation of nickel carbonyl is highly selective. In the process operated by International Nickel Ltd. at Clydach, South Wales, the granular nickel oxide feedstock is first reduced to impure metal with hydrogen in a rotary kiln; this is then fed directly into a similar volatiliser kiln in which the metal meets a counterflow of carbon monoxide at about 60°C. The evolved carbonyl gas goes to decomposers in which it passes over preheated nickel pellets on which it decomposes, depositing an additional thin layer of nickel on the pellets and freeing the carbon monoxide for recycling. The pellets are continuously circulated through the preheaters and decomposers, gradually growing in size until at about 8 mm diameter they are diverted by a sizing device as the finished product.

Nickel carbonyl is decomposed under different conditions to produce nickel powder. For this, a batch process is used in which purified liquid carbonyl is vaporised and fed to a reaction vessel heated to about 300°C. The sudden temperature rise decomposes the carbonyl to yield a powder with an average particle size in the order of 2–10 μm. The size range and other characteristics of the powder can be controlled by variation in the conditions of decomposition.

Hydrometallurgical processes. Hydrometallurgical processes are also used for the extraction of nickel from sulphide concentrates. In one such process the material is leached in aerated, strong ammonia at an elevated temperature and pressure (about 80°C and 10 atm). The leach solution is purified and converted to sulphate, and then reduced with hydrogen at about 200°C and 25–30 atm pressure. Nickel is deposited as powder which grows in a roughly spherical form and can be graded to different size fractions.

Lateritic ores

These ores are not susceptible to the physical processes of beneficiation used with sulphides, thus chemical treatment must be used to extract the nickel. Again both pyrometallurgical and hydrometallurgical methods are used. The first involves melting the ore with either sulphidation or reduction to enable a nickel–iron concentrate to be separated from the gangue. The second involves either selective reduction of the nickel oxide fraction, followed by ammoniacal leaching, or direct leaching of the ore with sulphuric acid at an elevated temperature and pressure. All, except the latter pressure-leaching process, initially require the ore to be dried, as it may contain up to 50 per cent free or combined water.

Pyrometallurgical processes. One of the pyrometallurgical processes used with oxide ores involves smelting the ore with a sulphiding material, usually gypsum (calcium sulphate). This produces a sulphide matte which can then be treated in a similar way to the matte derived directly from sulphide ores. A major production process using New Caledonian ores roasts the matte to remove sulphur. The oxide is then briquetted with a carbonaceous binder and reduced with charcoal in a muffle furnace, fired with producer gas, to a temperature around 1300°C. The metal sinters to form "rondelles", cylinders about 30 mm diameter and 15 mm high, with a purity of about 99.3 per cent. Alternatively, the ore may be smelted without the gypsum, using coke as a reducing agent and yielding ferro-nickel directly. The smelting is usually carried out in electric-arc furnaces.

Hydrometallurgical processes. The ammoniacal leaching process first requires the dried ore to be selectively reduced so that the nickel oxide is converted to nickel, the iron oxide remaining. This is effected by using producer gas under carefully controlled conditions at about 750°C. The product is leached with ammoniacal ammonium carbonate solution which takes nickel into solution, and from which it is precipitated as basic nickel carbonate by being heated. Nickel oxide is produced from this by kiln roasting, and may be marketed as such, or partially reduced and sintered to a form suitable for the steel industry.

The pressure acid-leaching process treats the lateritic ore directly. The mined material is screened and made into a water slurry, which is steam-heated in reaction autoclaves to about 250°C and treated with 98 per cent sulphuric acid. This takes the nickel and cobalt into solution, but very little of the iron. The free acid is neutralised, and the nickel and cobalt precipitated as sulphides by hydrogen sulphide. These are treated as the feedstock for a chemical refining process involving re-solution as sulphate, and hydrogen reduction at 190–200°C and 50 atm pressure, to yield a pure nickel powder.

FORMS OF NICKEL

Nickel as a raw material is available in a number of different forms suited to the variety of uses to which it is to be applied. The main forms for use in the metallurgical industries are listed in Table 7.

The higher-purity forms are used for applications in which the

TABLE 7

Market forms of primary nickel.

Form	*Approximate nickel content (%)*	*Principal use*
Electrolytic cathode	99.9	Alloys, plating
Carbonyl pellets	99.9	Alloys, plating
Rondelles	99.3	Alloys
Briquettes	99.9	Alloys
Carbonyl powder	99.9	Sintered parts, battery electrodes
Hydrogen-reduced powder	99.9	Porous and sintered parts
Ferro-nickel	20 upwards	Steel-making
Nickel-oxide sinters	75, 90	Steel-making

exclusion of extraneous elements is essential, particularly for alloys with special physical characteristics, e.g. magnetic materials, and for nickel-base superalloys. Carbonyl nickel has one particular merit in that its cobalt content is lower than that of other forms; this may be of significance in alloys for use in the in-pile regions of nuclear plant. Ferro-nickel and the nickel oxide sinters provide a source of nickel more economical than the above for steel-making and other ferrous alloy production.

Reference

1. Boldt, J. R., and Queneau, P., *The Winning of Nickel*, Methuen, London, 1967.

CHAPTER THREE

Nickel and Nickel Alloys

PURE NICKEL

It was emphasised in Chapter 1 that the purity and the previous metallurgical history of the material can have marked effects on some of its properties. The same is true in the case of many of the mechanical properties of the metal, and care must be taken to distinguish between the properties of high-purity metal (in the order of 99.99 per cent nickel) and commercial-purity nickel (say 99.5 per cent). Additionally, a range of wrought nickel products is available in which minor variations in composition are introduced in order to make it suitable for the particular application or processing procedure envisaged. In general, these materials contain about 99 per cent nickel. The effects of thermal history and working procedures can have a very marked influence on properties, such as impact strength, which are particularly structure-sensitive. These factors do, of course, influence the properties of all metallic materials, but they tend to be more critical with nearly-pure metals than they are with more highly-alloyed materials.

Composition and specifications

The forms of primary nickel available as the basis for the production of commercial alloys were listed in Table 7. The wrought alloys described as nickel embrace a number of materials with varying residual contents of deoxidants and other elements, some of which are given in Table 8, together with relevant British and American specifications. It is to be noted that most supplies of nickel, other than those refined by the carbonyl process, contain about 0.5 per cent of cobalt; in general this does not affect the properties of the metal, or of alloys made from it, and it has been normal commercial practice to express the nickel-plus-cobalt

TABLE 8

Nominal compositions and specifications for nickel wrought forms.

Material	*Nominal composition*							*Specifications**	
	Ni	*Cu*	*Fe*	*C*	*Mn*	*Si*	*Mg*	*BS*	*A.S.T.M.*†
Nickel 200	99.5	0.13	0.2	0.08	0.2	0.05	—	3072–3076 NA 11	B160–B163
Nickel 201	99.6	0.13	0.2	0.01	0.2	0.05	—	3072–3076 NA 12	B160–B163
Nickel 205	99.5	0.1	0.1	0.02	0.2	—	0.04	—	—
Nickel 270	99.98	—	—	0.01	—	—	—	—	—

* The specifications cover the following forms of product:

BS	*Form*	*A.S.T.M.*
3072	Plate	B162–61
3072	Sheet	B162–61
3073	Strip	B162–61
3074	Tube	B163–61
—	Pipe	B161–61
3076	Rod	B160–61
3076	Sections	B160–61
3075	Wire	—

† American Society for Testing Materials.

content as "nickel". The two principal types of material are "Nickel 200" (previously termed "A Nickel"), the general-purpose material, and "Nickel 201", the low-carbon version which is more suitable for applications involving much cold working and for use at elevated temperatures. The low carbon gives slightly higher ductility and avoids the danger of stress-corrosion cracking due to graphitisation at temperatures above about 300°C. The other grades of nickel are mainly of special interest to the electronic industry, as the trace elements influence the electron emission characteristics.

Mechanical properties

Hardness. Fully annealed high-purity nickel has been reported to have a hardness, as measured on the Vickers Diamond Pyramid scale, as low as 64 HV, but Nickel 200 and 201 have annealed hardnesses of about 90 and 75 HV respectively. The increase in the hardness of the two latter materials on cold working is illustrated in Fig. 11. Annealing is effected by heating to temperatures in the range 700–900°C, as indicated by the curves of Fig. 12, which embrace the temperatures normally used in practice.

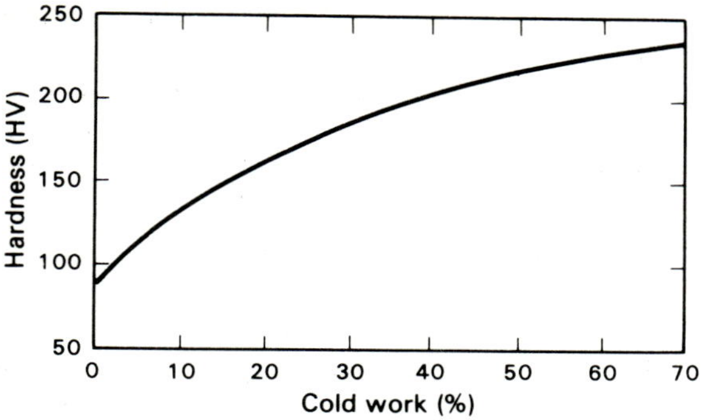

Fig. 11—Effect of cold work on the hardness of nickel.

The hardness of electrodeposited nickel is significantly higher than that of the wrought metal and depends greatly on the deposition conditions. For applications requiring high wear-resistance (for example, built-up worn steel parts), hardnesses up to 500 HV may be obtained. The increased hardness is partly due to the fine grain size of the electrodeposit and partly to co-deposited non-metallic materials. The topic is discussed in more detail in Chapter 6.

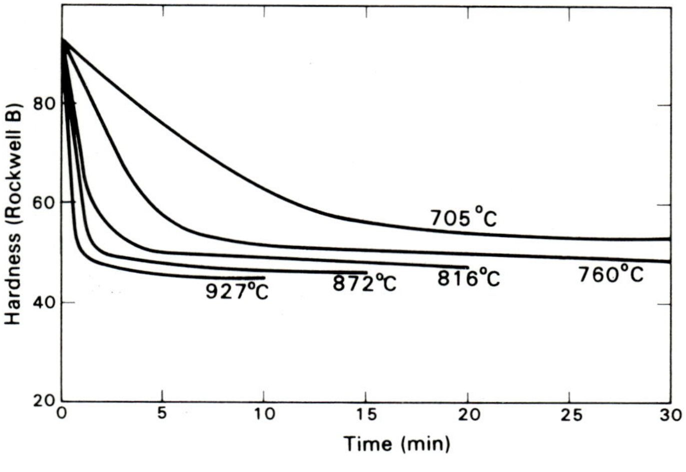

Fig. 12—Annealing curves for cold-worked nickel.

Impact properties. Nickel is an extremely tough metal and the high impact strength is maintained at sub-normal temperatures. The Izod impact strength of notched bars of Nickel 200 has been stated to be in the order of 160 J (120 ft lb) and to be little affected either by cold working to a hardness of over 170 HV, or by cooling to temperatures below 100 K.

Elastic moduli. Single crystals of nickel show anisotropic elasticity, the values of Young's modulus in the three simplest crystal directions being as follows:

$\langle 100 \rangle$	138×10^3 MN/m^2	(1.38×10^{12} dyn/cm^2)
$\langle 110 \rangle$	215×10^3 MN/m^2	(2.15×10^{12} dyn/cm^2)
$\langle 111 \rangle$	262×10^3 MN/m^2	(2.62×10^{12} dyn/cm^2)

Values for polycrystalline metal are a balance derived from these values and the measured modulus lies in the range 190–220 $\times$ 10^3 MN/m^2. It is probable that the value for completely randomly oriented crystals is close to 200 $\times$ 10^3 MN/m^2, measurements differing from this being due to a measure of preferred orientation in wrought material. In drawn wires or rods the $\langle 111 \rangle$ direction is the preferred fibre axis, while in rolled sheet there is a double structure (110), $\langle 1\bar{1}2 \rangle$; (112), $\langle 111 \rangle$. The proportions of the two structures depend on the conditions of rolling and annealing,

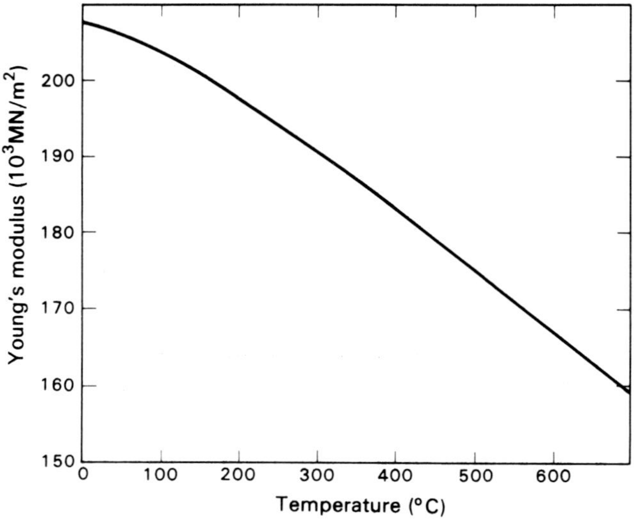

Fig. 13—Variation of Young's modulus of nickel with temperature.

but the first system generally predominates. The preferred orientation is not completely removed on recrystallisation, a new structure, predominantly (001), ⟨100⟩, replacing the original one.

The modulus falls with increasing temperature. The variation of the modulus of low-carbon nickel (Nickel 201) is shown in Fig. 13.

Some investigators have reported inflections in the Young's modulus in the region of the Curie point, and these appear to be associated with the degree of magnetisation, being at a

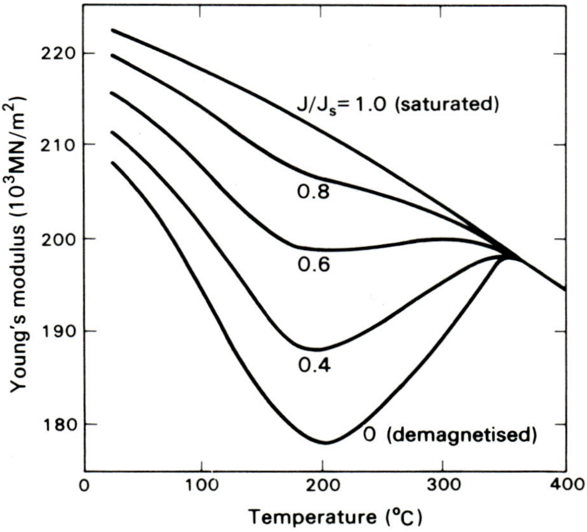

Fig. 14—Effect of magnetisation on Young's modulus of nickel.

maximum in the demagnetised condition but disappearing when the material is magnetically saturated. Figure 14 shows the effect of magnetisation on the modulus/temperature curve.

The torsional modulus or modulus of rigidity lies in the range $75–82 \times 10^3$ MN/m^2. Poisson's ratio, the ratio of lateral contraction to longitudinal extension for uniaxial stress, is 0.31.

Tensile properties. The tensile strength of nickel is in general closely related to the hardness. Figure 15 shows this relationship in strength and ductility parameters of Nickel 200. Depending on the degree of cold work and the annealing treatment, with the hardness ranging from 90 to 250 HV, the tensile strength at normal

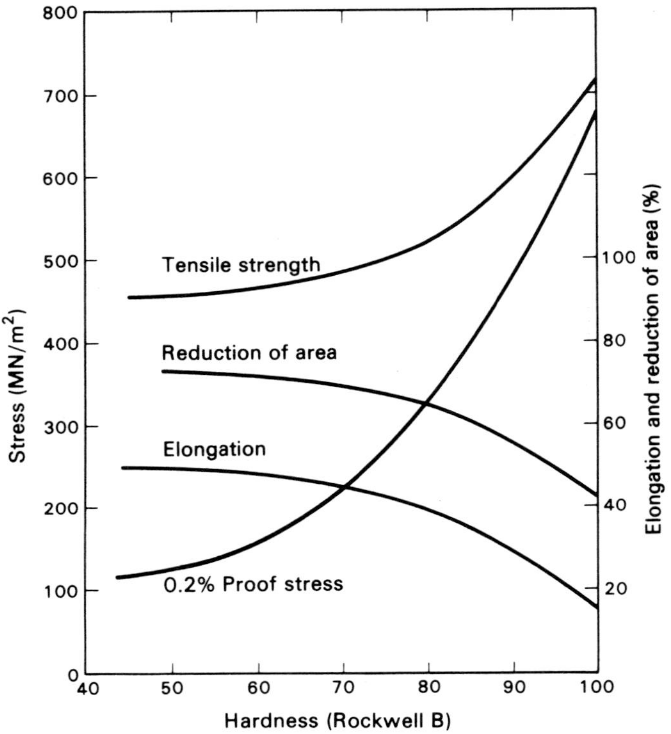

Fig. 15—Tensile properties of nickel in relation to hardness.

temperature can vary from 380 to 1100 MN/m². The low-carbon Nickel 201 can be annealed to give a somewhat lower tensile strength of 350 MN/m² with higher elongation, while high-purity nickel can give as low a tensile strength as 320 MN/m². As with the hardness, the strength falls with increasing temperature. Table 9 gives the tensile properties of hot-rolled and annealed Nickels 200 and 201 for temperatures up to 800°C. With decreasing temperature the tensile strength increases and Nickel 200 is reported to have a strength of 710 MN/m² at −195°C (78 K). It is to be noticed from Table 9 that the values of tensile elongation are not markedly affected by temperature rise up to about 350°C, and similarly the elongation is maintained at low temperatures—indeed the elongation of annealed nickel has been reported to be about 60 per cent, even close to absolute zero.

TABLE 9

Tensile properties of Nickel.

	Nickel 200			*Nickel 201*		
Temperature (°*C*)	*0.2% Y.S.* (MN/m^2)	*Tensile strength* (MN/m^2)	*Elong.* (%)	*0.2% Y.S.* (MN/m^2)	*Tensile strength* (MN/m^2)	*Elong.* (%)
20	148	462	47	—	403	50
100	153	459	46	141	383	45
200	140	459	44	121	372	44
300	138	462	46	124	365	42
400	115	328	64	95	303	—
500	100	245	68	86	245	—
600	69	176	72	76	176	—
700	59	121	82			
800	45	83	95			

Other mechanical properties. The other mechanical properties of nickel are perhaps even more structure-sensitive than the tensile properties and hence the values reported in the literature tend to be very variable. Nevertheless, some comments on the observations are merited. At temperatures up to about 300°C nickel does not show appreciable creep strain when stressed below the short-time tensile strength, but at higher temperatures normal creep behaviour is observed, i.e. a primary stage of decreasing creep-rate, a secondary stage of approximately constant creep-rate and finally a tertiary stage of accelerating creep leading to fracture. The rates of creep at any particular stress and temperature are very dependent on the composition, grain size and prior history of the material, but it may be taken as typical that high-purity nickel annealed at about 1000°C has a stress-rupture life, at 500°C under a stress of 30 MN/m^2, of about 10^5 min, which reduces to about 10 minutes under a stress of 120 MN/m^2. The minimum creep-rate of 99.75 per cent nickel, annealed to give a grain size of 0.033 mm, has been found to be proportional to the expression.

$$4.6\,e^{-Q/RT}$$

where the activation energy, Q, is 270 kJ/mol (65 000 cal/mol). This applies over the temperature range 400–1100°C and for stresses less than 70 MN/m^2, but increases more rapidly for higher stresses.

Cast nickel. Nickel is not extensively used in the cast form, but some components of chemical plant, for which corrosion-resisting qualities are required, are conveniently made by sand casting. A silicon content of about 1–2 per cent and a manganese content of about 1 per cent would then be suitable and the melt should be deoxidised with magnesium, leaving a residue of about 0.1 per cent. The properties of such material are then as follows:

hardness	100–125 HV
0.1% proof stress	120 MN/m^2
tensile strength	360–420 MN/m^2
elongation	10–30 %

Other nickel-rich alloys. A number of commercial grades of nickel contain small percentages of other elements which are added to improve particular characteristics, but which do not change very markedly the general properties of the material. Thus additions of manganese up to about 5 per cent confer much improved resistance to sulphur attack at elevated temperatures, since MnS

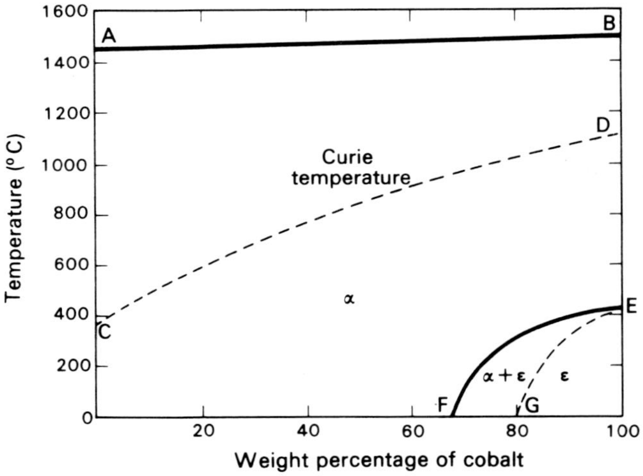

Point	A	B	C	D	E	F	G
°C	1453	1495	353	1121	417	0	0
%Co	0	100	0	100	100	68	80

Fig. 16—Nickel–cobalt equilibrium diagram.

is formed in preference to Ni_3S_2, and this avoids the low-melting point eutectic referred to in Chapter 1, p. 16. The manganese is in solid solution in the nickel and improves the strength and hardness without appreciable reduction of ductility. Such alloys are used for electrode support wires in radio valves and tungsten-filament lamps. Similar alloys, frequently with the silicon content increased to about one per cent, are used for the electrodes of sparking plugs.

Small cobalt additions in the region of 4–5 per cent are sometimes made to nickel to improve its magnetostrictive properties. Nickel forms a complete range of solid solutions with face-centred-cubic cobalt, the hexagonal structure being restricted to alloys containing less than about 30 per cent nickel at normal temperature. Figure 16 shows the diagram of the nickel–cobalt system[1]. The effects of cobalt additions to nickel are to increase the Curie point and the magnetic saturation, and to slightly improve the mechanical properties. Nickel–cobalt alloys containing up to about 30 per cent cobalt are formed by electrodeposition from mixed-salt electrolytes, and they have an exceptionally high hardness (*see* Chapter 6).

NICKEL–COPPER ALLOYS

A complete range of solid solutions is formed between nickel and copper with no evidence of ordering; the constitutional diagram is shown in Fig. 17. The Curie point falls with increasing copper content until it reaches normal temperature at about 30 weight per cent copper. The electrical resistivity of the binary alloys increases steeply at each end of the diagram and reaches a maximum about thirty times that of pure copper and four times that of pure nickel at close to 50 atomic per cent. Alloys of industrial importance are based on both copper-rich and nickel-rich binary alloys, and to some extent the choice of composition is determined by economic factors—nickel costs two to three times as much as copper.

Cupro-nickel alloys

These are copper-base alloys but nevertheless merit a short note because of their importance in the markets for nickel. Alloys containing 10–30 per cent nickel have outstanding corrosion-resisting properties, particularly to sea-water, coupled with progressively improved strength over that of pure copper (*see* Table 10). Their hot and cold workability is still such that they are readily fabri-

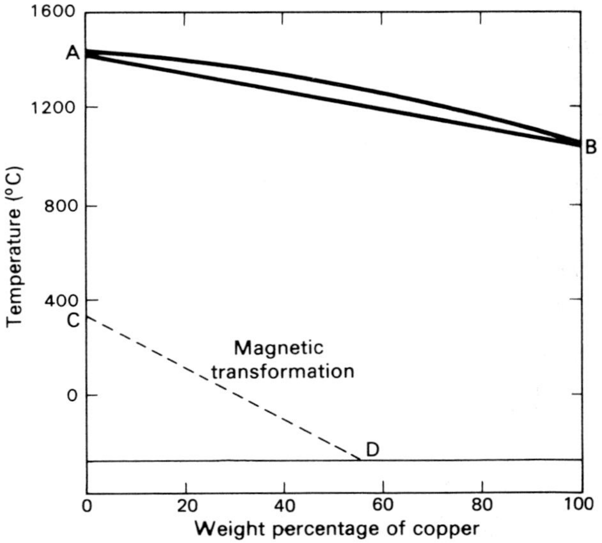

Point	A	B	C	D
°*C*	1453	1083	353	−273
%*Cu*	0	100	0	58

Fig. 17—Nickel–copper equilibrium diagram.

cated into the required shapes and drawn into seamless tubing, in which form they are widely used for heat-exchanger equipment in marine engineering. In this field, alloys containing 10 or 30 per cent of nickel are common. The alloys work-harden fairly rapidly, but by repeated working and annealing it is possible to

TABLE 10

Mechanical properties of annealed wrought copper–nickel alloys.

Alloy *Cu*	*Ni*	*Tensile strength* (MN/m^2)	*Elongation* (%)	*Hardness* (*HV*)
95	5	275	45	75
90	10	320	50	80
80	20	340	65	90
70	30	370	50	95

TABLE 11

Properties of work-hardened copper–nickel alloys.

Alloy		*Annealed*		*Half-hard*		*Hard*	
Cu	*Ni*	*Hardness*	*Tensile strength*	*Hardness*	*Tensile strength*	*Hardness*	*Tensile strength*
(%)		(*HV*)	(*MN*/*m*²)	(*HV*)	(*MN*/*m*²)	(*HV*)	(*MN*/*m*²)
90	10	77	320	137	410	150	480
70	30	93	370	154	480	174	560

develop complex shapes. The increased strength obtained in the cold-worked condition with sheet material is shown in Table 11. The recommended temperatures for hot working are 840–950°C for the 90/10 alloy and 920–1020°C for the 70/30 alloy, while the annealing ranges are 595–820°C and 650–820°C respectively. The alloys can be welded by a variety of processes, details of which are given in Chapter 6, p. 111. Although by far the greatest proportion of these alloys is used in the wrought form, castings are also made, and provided control of extraneous elements is adequate (for example, lead should not exceed 0.01 per cent), these can be welded by appropriate techniques to other castings and to wrought forms.

Cupro-nickel alloys normally contain controlled amounts of iron, about 1.5 per cent for the 90/10 alloy and about 0.5 per cent for the 90/30 alloy, since this is found to improve the corrosion performance in moving sea-water. The corrosion rates of such alloys in quiet sea-water can be compared with that of copper thus:

Criterion in quiet sea-water (velocity 1 m/s)	Cu	*Alloy* 90 Cu/10 Ni	70 Cu/30 Ni
Average corrosion rate (μm/year)	10–100	2.5–10	2.5–10
Typical depth of pitting (μm/year)	150–300	25–130	25–130

In more rapidly moving sea-water, the resistance of a material to corrosion may be markedly increased. While copper may corrode at up to 75 μm/year at velocities not exceeding 0·6 m/s, the cupro-nickels corrode at less than 25 μm/year at water speeds up to 2.5 m/s for the 90/10 alloy and up to 5 m/s for the 70/30 alloy.

"Nickel-silver" is a term used to describe a wide range of copper–nickel–zinc alloys, originally used mainly for decorative applications as an economic replacement for silver. Although the alloys are still widely used for ornamental purposes, both uncoated and as a base for silver plating, they also have attractive mechanical and corrosion-resisting qualities that make them useful for a range of engineering applications. The alloys are copper-base with contents of 10–25 per cent nickel and 15–45 per cent zinc. Above about 12 per cent nickel the yellow colour of copper–zinc brass is completely suppressed and the alloys are white.

There are two main classes of nickel-silver, the single-phase "alpha" alloys containing below about 37 per cent zinc (hence most suitable for cold-working operations and the development of work-hardened properties), and the two-phase "alpha-beta" alloys (more suitable for hot-working operations, particularly extrusion). It is not possible here to quote the full range of properties obtainable with nickel-silvers, but by appropriate working and annealing, hardnesses ranging from below 100 HV to over 250 HV are obtainable.

The characteristics which make them attractive are as follows:

1. Resistance to corrosion and tarnishing.
2. Resistance to stress-corrosion, particularly the lower zinc-content alloys.
3. Good strength and ductility.
4. Good cold workability.
5. Work hardens to spring temper.
6. Non-magnetic.

The wide range of applications of nickel-silvers is dealt with in Chapter 7.

Nickel–copper alloys

In the early years of the present century attempts were made to produce a useful alloy direct from the mixed nickel–copper sulphides of the Sudbury deposit, without separation and refining of the two metals. The attempt was successful, and the alloy known as "Monel", containing about 70 per cent nickel and 30 per cent copper, was developed. Now termed Monel alloy 400, it is no longer made directly from the sulphide matte, but is synthesised from the refined metals. It is characterised by good strength and ductility, and resistance to the corrosive action of a wide range of media—particularly the atmosphere and sea-water.

The nominal composition of Monel alloy 400 and its British

Standards and American Society for Testing Materials (A.S.T.M.) specifications are given below:

Cu	*Fe*	*Mn*	*C*	*Ni*
		(as percentage)		
30	2.5 max.	1	0.15	bal.

Specifications

BS	*A.S.T.M.*
3072–3076: NA 13	B127, B163–165,
3071: NA 1–3 (cast)	B395

It is available in all the common wrought-forms and also as castings. Typical mechanical properties of the wrought alloy in the hot-rolled and annealed condition are as follows:

0.2% proof stress (MN/m^2)	*Tensile strength* (MN/m^2)	*Elongation* (%)	*Hardness* (*HV*)
200	550	48	125

The corresponding properties of the cast alloys, of which versions exist with silicon contents ranging from 1–4 per cent, fall within the ranges given below. (The low-silicon alloys complement the wrought alloys, and the high-silicon alloys have improved resistance to erosion by steam and water.)

0.1% proof stress (MN/m^2)	*Tensile strength* (MN/m^2)	*Elongation* (%)
150–240	430–620	10–16

The strength of the alloy with about 4 per cent silicon can be increased by heat treatment from about 260 HV in the cast condition to 300–380 HV after heating to 595°C for 6 hours and air cooling.

Alloy 400 is not susceptible to heat treatment to improve its mechanical properties, but its strength can be improved by cold working. In the cold-drawn and stress-relieved condition, the properties can fall within the following range:

0.2% proof stress (MN/m^2)	*Tensile strength* (MN/m^2)	*Elongation* (%)	*Hardness* (*HV*)
380–700	580–830	35–22	160–220

The strength of the annealed alloy is retained at temperatures up to about 540°C but then falls with increasing temperature, until at 600°C it has fallen to about half its normal-temperature value. It has the best high-temperature strength of the binary nickel–copper alloys and also has the best creep strength, although this is much lower than that of the special creep-resisting alloys based on the nickel–chromium system.

The resistance of an alloy to corrosion normally depends on many factors, which makes it difficult to express behaviour in reliable quantitative terms meaningful in other than carefully-controlled laboratory conditions. Nevertheless, tests made under such conditions, coupled with long practical experience, enables a reasonable picture to be formulated. Monel alloy 400 has been shown to be suitable for use in fresh water of all types, sea-water, brine, superheated steam, non-oxidising acids, neutral and alkaline salts, alkalis, ammonium sulphate and most pharmaceutical products. It is not resistant to oxidising acids such as HNO_3, HNO_2, H_2SO_4 and hypochlorites.

The resistance to acids depends greatly on the degree of aeration, which in all cases markedly increases the rate of attack. In systems purged of air by nitrogen, the rates of attack by mineral acids are usually less than about 250 μm/year, as shown by the following examples:

1. In sulphuric acid of concentrations up to 85 per cent at 30°C, the rate of attack does not exceed 180 μm/year; at higher concentrations the rate of attack sharply increases.
2. In phosphoric acid at all concentrations the rate of attack is less than 50 μm/year up to 50°C and less than 250 μm/year up to 100°C.
3. In hydrofluoric acid the attack is little affected by acid concentration or temperature, not exceeding about 200 μm/year.

In quiet sea-water the average rate of attack on Monel alloy 400 is usually less than 25 μm/year, although pitting up to about 400 μm/year may occur—in this latter respect the alloy is less resistant than the cupro-nickels. In faster-moving sea-water, say over 2 m/s, pitting is avoided and the average corrosion-rate remains within the same range.

Nickel–copper alloys can be made age-hardenable by the inclusion of small amounts of aluminium and titanium, and after appropriate heat treatment, sometimes combined with cold work, high strengths can be achieved. Monel alloy K-500 contains about 3 per cent aluminium and 0.5 per cent titanium. Precipitation hardening is normally effected by ageing for 16 hours at about 580°C after hot working. Typical properties then obtained are as follows:

0.2% proof stress (MN/m^2)	*Tensile strength* (MN/m^2)	*Elongation* (%)	*Hardness* (*HV*)
760	1100	20	300

By ageing the cold-worked alloy, the tensile strength can be increased to about 1200 MN/m^2 and the hardness to about 320 HV. The high strength, corrosion resistance and non-magnetic properties make the alloy suitable for non-sparking tools, marine propeller-shafts, etc.

Nickel–copper alloys for electrical uses

Alloys containing approximately equal atomic proportions of the two metals (48 weight per cent nickel) have a high electrical resistivity and low temperature coefficient of resistivity. Because of their ductility and ease of working, they are widely used for resistance wires which are not required to operate at temperatures exceeding about 400°C. For higher temperatures the nickel–chromium alloys are necessary (*see* p. 42).

The alloy containing 46 weight per cent nickel, known as Constantan or Ferry, is widely used; it has a resistivity at 20°C of $49 \times 10^{-8}\,\Omega m$ and a very low temperature coefficient. The coefficient is dependent on exact composition and treatment but can be as low as 0.000 02 per deg. C over the temperature range 0–100°C. The same alloy is also used as a thermocouple element since it has a high thermoelectric power. The e.m.f.s developed against platinum, copper and iron, for junctions at 0 and 100°C, are as follows:

	Pt	*Cu*	*Fe*
e.m.f. (*mV*)	3.48	4.24	5.26

Nickel–copper alloys containing around 70 per cent nickel have also been used as magnetic temperature-compensating alloys, since the permeability changes rapidly with temperature in the region of the Curie point—it will be recalled that at around this composition the Curie point is near to ambient temperature.

NICKEL–CHROMIUM ALLOYS

The constitutional diagram of the nickel–chromium system is

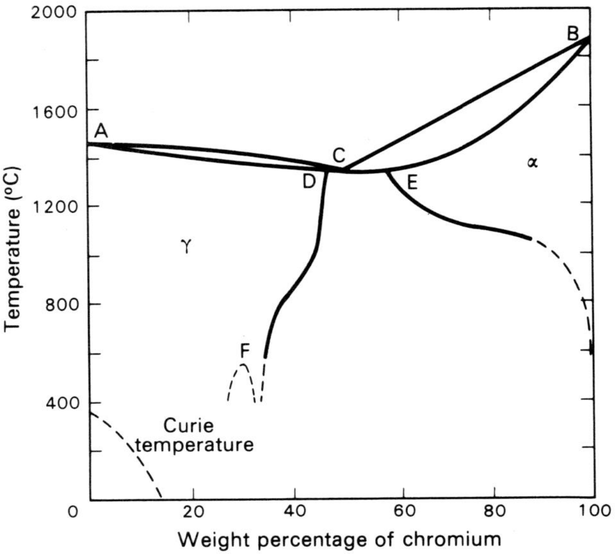

Point	A	B	C	D	E	F
°C	1453	1880	1345	1345	1345	580
%Cr	0	100	51	47	59	31

Fig. 18—Nickel–chromium equilibrium diagram.

shown in Fig. 18. Chromium is taken into solid solution in nickel to the extent of about 30 per cent at normal temperature, and the principal effects of increasing chromium are to modify the oxidation resistance of the nickel, to increase the strength, and to increase the electrical resistivity. These effects on the properties of nickel are all exploited in various alloys based on the simple binary system.

Electrical resistance heating elements

The influence of the chromium content on the oxidation resistance of nickel is illustrated by Fig. 19. This shows that, in agreement with Wagner's theory of the effect of introducing ions with different valency into the scale, small additions of chromium initially increase the rate of oxidation by increasing diffusion rates in the scale. However, beyond about 7 per cent chromium in the alloy, the constitution of the scale changes significantly and much improved oxidation resistance is shown, increasing progressively with increasing chromium content. Substantial increase of electrical resistivity and progressive increase in strength with reduction in ductility are found with increasing chromium content, and the outcome of these changing properties is that the binary alloy containing about 20 per cent chromium provides the optimum balance for an alloy to be drawn to wire and used for electrical heating elements at high temperatures. The alloy, known under various trade names such as Nichrome or Brightray, is capable of long service life at temperatures up to 1150°C. Various modifications in the contents of the minor constituents are made to suit the particular conditions of use. The resistance to progressive

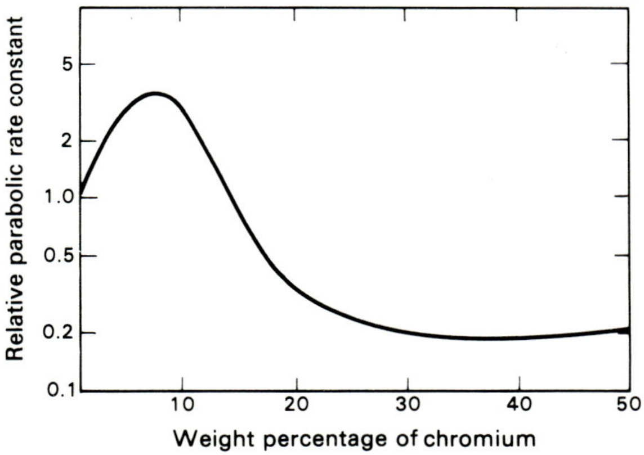

Fig. 19—Relative oxidation resistance of nickel–chromium binary alloys.

oxidation at high temperatures is due to the formation of an adherent protective scale. The action of the small amounts of reactive elements, silicon, calcium, cerium, zirconium, etc., is to improve the coherence and adherence of this scale. The scale itself consists of a mixture of nickel oxide (NiO) and chromium oxide (Cr_2O_3) which partially combine to form the complex oxide, nickel chromite ($NiCr_2O_4$), with a spinel structure. It is, however, by no means clear whether the oxidation resistance depends to a significant extent on the formation of the spinel. On the other hand, it is well established that the adhesion of the scale to the metal is much improved by appropriate additions of the reactive elements mentioned above. For intermittent use, in which the repeated heating and cooling of the elements tend to crack and flake the scale from the metal, a typical composition of the alloy would be as follows:

Cr	*Si*	*Ca*	*Ce*	*Ni*
		(by percentage)		
20	1.5	0.1	0.05	bal.

For continuous use, such as in industrial furnaces, where scale flaking is less likely to occur, lower levels of the reactive elements are suitable, and a typical composition would be:

Cr	*Si*	*Ca*	*Ni*
	(by percentage)		
20	0.5	0.05	bal.

The electrical and mechanical properties of the alloy are not significantly affected by the compositional differences, typical properties being shown in Table 12.

TABLE 12

Typical tensile properties of Nichrome or Brightray alloy.

Temperature (°*C*)	*Tensile strength* (MN/m^2)	*Elongation* (%)
20	740	47
400	620	46
500	600	46
600	500	27
700	390	26
800	210	36
900	120	43
1000	70	68

Electrical resistivity at 20°C: 115×10^{-8} Ωm.
Temperature coefficient of resistivity: 0.000 08 per °C.

At about 500°C there is an inflection in the resistivity/temperature curve shown in Fig. 20, the properties on cooling depending on the cooling rate. It has been suggested that this phenomenon can be explained on the basis of ordering, based on the stoichio-

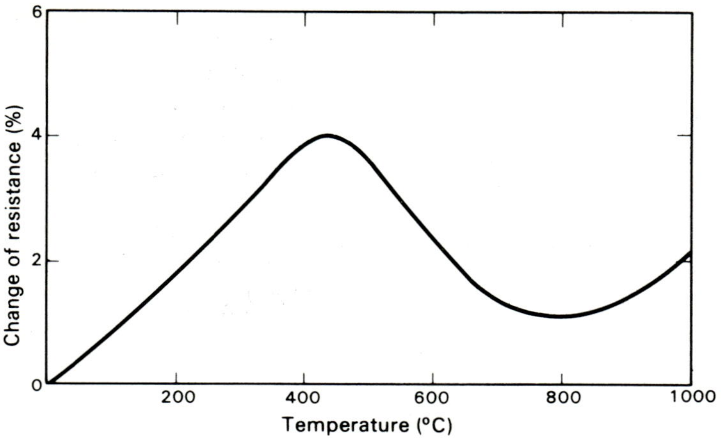

Fig. 20—Change of resistance with temperature of 80/20 nickel–chromium alloy.

metric formula Ni_3Cr, but this cannot be confirmed by X-ray detection of the superlattice because of the similarity in scattering factors of nickel and chromium.

The binary alloy containing 10 per cent chromium is also used as an electrical heating element. However, its major use is as one element of a thermocouple, the other element being a 95 nickel, 5 aluminium alloy. This couple, termed Chromel-Alumel, is widely used for industrial applications at temperatures up to about 1100°C.

Hot corrosion of nickel–chromium alloys

Apart from their use in the form of wire for electrical purposes, nickel–chromium alloys, particularly the 20 per cent chromium alloy, are used as wrought parts and castings for high-temperature purposes. They are more resistant to oxidation and most forms of hot corrosion than the cheaper iron–nickel–chromium alloys (*see* pp. 47–50).

For pure oxidation resistance there is rarely need to consider alloys with more than 20 per cent chromium, but for corrosion by fuel ashes and other deposits involving alkali–metal salts, particularly sulphates, 'there is marked benefit to be obtained by using much increased chromium contents. Hot corrosion of this nature is a complicated topic and depends on the details of the composition of the ash as well as the temperatures involved. There are, however, two main mechanisms of attack to be considered.

The first is accelerated oxidation arising from reactions between the ash constituents and the normal oxide scale on the metal. Attack of this nature is particularly promoted by vanadium-containing ashes such as arise from the combustion of some mineral oils. Vanadium pentoxide melts at about 660°C, and although it may be combined with other oxides to form higher-melting vanadates, it nevertheless has a fluxing action on the normal nickel–chromium scale and permits oxidation to proceed at an accelerated rate. It has also been suggested that vanadium pentoxide may act as a catalytic oxidant.

The second major mechanism is sulphidation arising from the presence of sulphates in the ash or of sulphur in the surrounding atmosphere. Chromium sulphide (Cr_2S_3), with a melting point

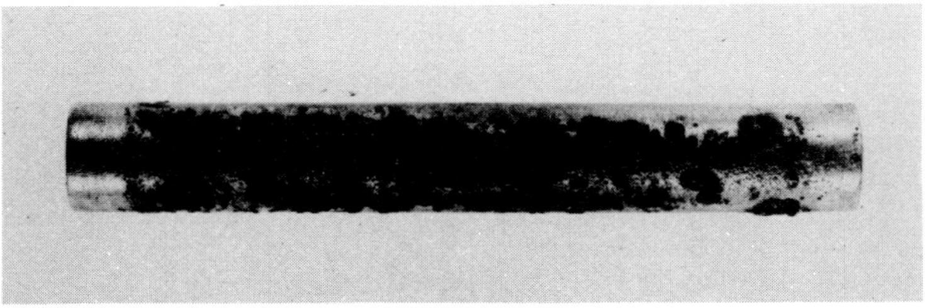

[*Inco Europe Ltd.*

Fig. 21—Sulphur corrosion at high temperature of nickel–chromium alloy.

in the order of 1550°C, is formed preferentially to nickel sulphide and hence delays the formation of the low-melting-point nickel/nickel-sulphide eutectic referred to in Chapter 1. Eventually, however, local concentration of sulphur-containing compounds can lead to exhaustion of the chromium and the formation of the eutectic, with subsequent rapid liquid-phase attack—this appears as characteristic wart-like growths of corrosion product (*see* Fig. 21). The behaviour of binary nickel–chromium alloys containing 20–60 per cent of chromium has been studied under simulated ash-corrosion conditions. With the most corrosive mixture used ($80\%\ V_2O_5 + 20\%\ Na_2SO_4$) the 60 per cent chromium alloy was shown to be the most generally resistant. Binary alloys containing more than about 30 per cent chromium have a two-phase structure consisting of γ-nickel and α-chromium solid solutions, and, because of the brittleness of α-chromium, the ductility of the alloy decreases with increasing chromium content. The normal-temperature tensile properties of cast binary alloys are as follows:

Chromium (%)	*Tensile strength* (MN/m^2)	*Elongation* (%)
35	480	62
50	540–680	7–24
60	800–1000	1–2

Alloys containing up to about 35 per cent chromium are sufficiently ductile to be hot worked, but the higher-chromium alloys are normally only suitable for casting, although ductility can be obtained by the addition of small quantities of zirconium.

The good oxidation resistance of nickel–chromium alloys coupled with their reasonable strength at elevated temperatures led to their forming the basis from which nickel-base creep-resisting alloys were developed—these are dealt with on pp. 51–61.

NICKEL–IRON–CHROMIUM ALLOYS

The ternary constitutional diagram for the system shown in Fig. 22 indicates a very wide region covered by the face-centred-cubic structure of nickel. It has already been seen that in the binary system up to about 30 per cent chromium is soluble in nickel. Figure 22 shows that the same phase-field stretches close to the iron corner of the diagram—indeed the familiar austenitic steels (containing 16–25 per cent chromium and 5–20 per cent nickel) belong to this field. Within the broad field there are a number of nickel-rich alloys of considerable industrial importance, mainly characterised by good corrosion resistance conferred by the chromium content, and good retention of mechanical strength at elevated temperatures. Binary nickel–iron alloys also within the same phase-field are dealt with on pp. 64–75.

There are two main groups of alloy to be considered. The first, termed Inconel alloys, are based on the composition 76 per cent nickel, 7 per cent iron and 16 per cent chromium, and were originally developed by introducing chromium into nickel by the most economical source, ferro-chromium. The basic alloy, now termed Inconel alloy 600, was originally used for the dairy industry because of its resistance to corrosion by milk, but has now largely been superseded for this purpose by austenitic steels. It has, however, good corrosion resistance to many media, both acid and

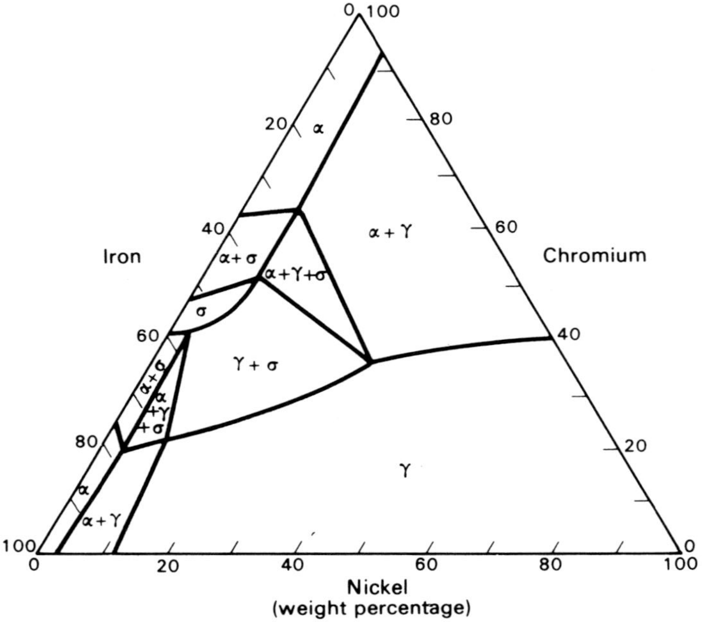

Fig. 22—Iron–nickel–chromium equilibrium diagram at 650°C.

alkaline, oxidising and reducing. It also has much of the high-temperature oxidation resistance of the binary nickel–chromium alloys already described, and hence is used widely for furnace parts and heat-treatment equipment. Mechanical properties of hot-rolled bar at normal and elevated temperatures are given in Table 13. The alloy retains reasonable strength at elevated tem-

TABLE 13

Mechanical properties of annealed Inconel alloy 600.

Temperature (°C)	*0.2% P.S. (MN/m²)*	*Tensile strength (MN/m²)*	*Elongation (%)*	*Hardness (HV)*
20	185–300	540–650	35–55	140–160
400	200	600	49	—
500	195	580	48	—
600	190	520	42	—
700	160	320	38	—
800	95	150	58	—

peratures and also has appreciable creep resistance at temperatures up to about 600°C.

In the same way as the binary nickel–chromium alloys have been developed by the addition of elements to confer precipitation-hardening properties to give rise to a family of special creep-resisting alloys, the Inconel alloys have led to a similar family—these are referred to on pp. 51–61. Additionally, however, other modifications have been made to the basic alloy to yield alloys with individual characteristics. Perhaps the most interesting is Inconel alloy 625 with increased chromium content and molybdenum and niobium additions in solid solution, giving the following nominal composition:

Cr	*Mo*	*Nb*	*Fe*	*C*	*Ni*
		(*by percentage*)			
22	9	4	5	0.05	bal.

This alloy is considerably stronger than Inconel alloy 600, retains its strength better at elevated temperatures, but has, nevertheless, better ductility (*see* Table 14). Both of the Inconel alloys mentioned are produced in most of the standard wrought forms.

TABLE 14

Mechanical properties of annealed Inconel alloy 625.

Temperature (°*C*)	*0.2% P.S.* (*MN/m²*)	*Tensile strength* (*MN/m²*)	*Elongation* (%)	*Hardness* (*HV*)
20	600	980	42	280
400	480	910	42	—
500	490	900	42	—
600	490	850	50	—
700	470	690	68	—
800	400	420	85	—

Alloys of the second group are termed Incoloy alloys, the basic composition being 32 per cent nickel, 20 per cent chromium, balance iron. Their main characteristic is resistance to hot corrosion, particularly in the field of industrial furnace practice where combined oxidation and carburisation are prevalent. The higher iron content increases the solubility for carbon so that there is less danger of the formation of embrittling chromium carbides than is the case with alloys richer in nickel. The alloy is far enough from the sigma field in the ternary diagram (*see* Fig. 22) to be immune to the formation of this intermetallic compound, unlike austenitic steels with lower nickel contents. There are three alloys of importance in this group, their nominal compositions being given in Table 15.

The strengths of these alloys are generally similar, annealed hot-rolled bar having a hardness of 150–200 HV, a tensile strength of 600–720 MN/m^2 and an elongation of 40–50 per cent. The main individual characteristics lie in their particular corrosion resisting behaviour. Incoloy alloy 800 is mainly of interest for resisting high-temperature oxidation and carburisation, but also has good resistance to wet corrosion in oxidising media. It is relatively free from stress-corrosion cracking, especially in chloride and hydroxide solutions, and significantly better in this respect than austenitic

TABLE 15

Nominal compositions (per cent) of three alloys of the Incoloy group.

Alloy	*Ni*	*Cr*	*Mo*	*Cu*	*Ti*	*Si*	*Fe*
Incoloy alloy 800	32	20	—	—	—	0.5	bal.
Incoloy alloy 825	40	21	3	2	1	0.5	bal.
Incoloy alloy DS	37	18	—	—	—	2.3	bal.

steels. However, alloy 825 is still more resistant under these conditions, and is particularly resistant to general attack by severely aggressive solutions such as strong mineral acids. At temperatures up to 50°C the corrosion rate in sulphuric acid of all concentrations does not exceed 125 μm/year. In boiling acids up to 40 per cent by weight the rate does not exceed 500 μm/year. It is also reasonably resistant to nitric acid, the corrosion rate being less than 25 μm/year at normal temperature for acid of all concentrations, and less than 125 μm/year at the boiling point of all acids up to 40 per cent concentration. Incoloy alloy DS is resistant to carburising conditions in furnaces at temperatures up to about 1050°C, and in particular minimises the danger of "green-rot". This is a type of attack observed with nickel–chromium alloys under cyclic oxidising and carburising conditions, characterised by the formation of chromium carbides and followed by oxidation, either of the carbide particles or the adjacent chromium-depleted matrix. The alloy becomes embrittled and the fractured surfaces have a green appearance.

NICKEL-BASE CREEP-RESISTING ALLOYS

Nickel-rich alloys based on the nickel–chromium binary alloys form the largest and most important group of alloys specially developed for service under highly-stressed conditions at high temperatures[2,3]. The primary mechanical requirement for such service is high creep resistance, and alloys with this characteristic are now generally termed "superalloys".

Structure and creep resistance

The metallurgical and structural factors leading to high creep resistance are reasonably well understood and a short outline of them is necessary before describing the alloys themselves.

Since creep is due to the thermally-activated movement of dislocations through the crystal lattice, the initial requirement for high creep resistance is a matrix which has a high elastic modulus and low diffusion rates at elevated temperatures. Although nickel is not exceptional in these respects, it has a wide tolerance for other elements in solid solution. Nickel-base superalloys commonly have contents of one or more of the elements chromium, cobalt, iron, molybdenum, tungsten and niobium. Chromium is an essential constituent to confer corrosion resistance on the alloy, and is normally present to the extent of 15–30 per cent; the other elements, if present, are usually at a lower level, varying from 3 to 20 per cent. The effectiveness of these solutes in strengthening the matrix depends on the difference in atomic size between nickel and the solute; the effect of the solute on the stacking fault energy; and the diffusion rate of the solute in nickel. A complex matrix containing several solutes is often found to be superior to one containing fewer solutes.

The second major strengthening mechanism is the presence of a stable, finely-dispersed second phase, most nickel-base superalloys relying on the intermetallic compound $Ni_3(Al, Ti)$, termed γ'. This phase has an ordered face-centred-cubic structure isomorphous with the matrix, and with a lattice parameter within one per cent of that of the matrix. The structural similarity allows the γ' phase to precipitate coherently with the matrix, giving great stability at elevated temperatures. The levels of titanium and aluminium necessary to permit formation of γ' are about 1–5 per cent of each element, and typically the total of the two is about 5–6 per cent. Other dispersed phases, particularly carbides, can contribute to lattice strengthening of superalloys, but the major significance of carbides is in relation to their influence on grain-boundary structure; there they control to a large extent the onset of intergranular fracture. The constitution and morphology of grain-boundary carbides depends not only on the major composition of the alloy and heat treatment applied, but also on the content of minor constituents such as boron and zirconium, present at levels of about 0.01 per cent.

COMPOSITIONS

There is a wide range of nickel-base superalloys available having properties making them suitable for service at elevated temperatures, in some cases up to 1100°C. They also have characteristics, either of mechanical properties, fabricability or corrosion resistance, which fit them to particular applications. The compositions of a few selected alloys used in the wrought form are given in Table 16. It is to be noted that the first two alloys listed, Nimonic alloy 75 and Hastelloy X, do not have aluminium and titanium contents high enough to precipitate γ'—they are essentially solid-

TABLE 16

Wrought nickel-base superalloys.

Alloy	*Cr*	*C*	*Al*	*Ti*	*Co*	*Mo*	*Nb*	*Fe*	*Ni*
Nimonic 75	20	0.1	—	0.4	—	—	—	—	bal.
Hastelloy X	22	0.1	—	—	1.5	9	—	18	bal.
Nimonic 80A	20	0.05	1.3	2.3	—	—	—	—	bal.
Nimonic 90	20	0.05	1.5	2.5	17	—	—	—	bal.
Nimonic 115	15	0.15	5.0	4.0	15	3.5	—	—	bal.
Inconel X750	15	0.04	0.9	2.5	—	—	0·9	7	bal.
Nimonic 901	12.5	0.5	0.2	2.9	—	5·7	—	bal.	42

solution-hardened alloys used mainly in the form of sheet because of their relatively high ductility and formability. The other alloys listed are all hardened by γ' precipitation; alloy 901 is basically a nickel–iron–chromium alloy and is capable of γ' hardening in the same manner as the nickel-base alloys. The alloys are mainly provided as bars or forgings, but some, particularly the more lowly-alloyed ones, are also produced as extruded and drawn tube.

For use as castings, somewhat different alloys are generally preferred, and typical compositions are given in Table 17. Nimocast 80 is equivalent in composition to the wrought alloy Nimonic 80A, but the aluminium and titanium contents make casting in air difficult because of their high affinity for oxygen. Alloy PE 10,

TABLE 17

Cast nickel-base superalloys.

Alloy	*Type*	*Cr*	*C*	*Al*	*Ti*	*Co*	*Mo*	*Nb*	*W*	*Ni*
Nimocast 80	A	20	0.05	1.3	2.3	—	—	—	—	bal.
Nimocast PE10	A	20	—	—	—	—	6.0	6.5	2.5	bal.
Nimocast 713	V	13	—	6·0	1·0	—	4.5	2.3	—	bal.
IN 100	V	10	0.18	5.5	4.7	15	3.0	—	—	bal.

A=air casting; V=vacuum casting.

hardened by niobium, is much more readily cast in air and has improved high-temperature properties. By using vacuum melting and casting techniques the full potentiality of hardening by γ' precipitation can be exploited. Nimocast 713 and alloy IN 100 are alloys of this type.

Mechanical properties

The creep and stress-rupture properties of superalloys are those of major significance in relation to their use at elevated temperatures, and they will therefore be considered first.

To give a full account of the creep behaviour of an alloy it is necessary to determine the dependence of the creep strain and the point of fracture on the three variables of stress, temperature and time. For this purpose a large number of creep tests under varying conditions must be made. Correlation of the results of such tests to enable interpolation and extrapolation to be made for other than the actual test conditions is frequently carried out by the use of empirical formulae, of which the best known is that of Larson and Miller[4]. In this formula a time–temperature parameter of the following form is calculated:

$$P = T\,(20 + \log t),$$

where T is temperature in kelvins and t is time in hours. The stress to produce specific values of creep strain, e.g. 0.1, 0.2, 0.5 per cent, or fracture, is then plotted against this parameter to produce a family of curves of the type shown in Fig. 23. These provide reasonable indications of the behaviour of the alloy over the range of conditions experimentally covered, and may be extrapolated to some extent without undue risk. Such curves are of particular value for comparing the behaviour of different alloys. Larson–Miller curves for the rupture properties of the alloys

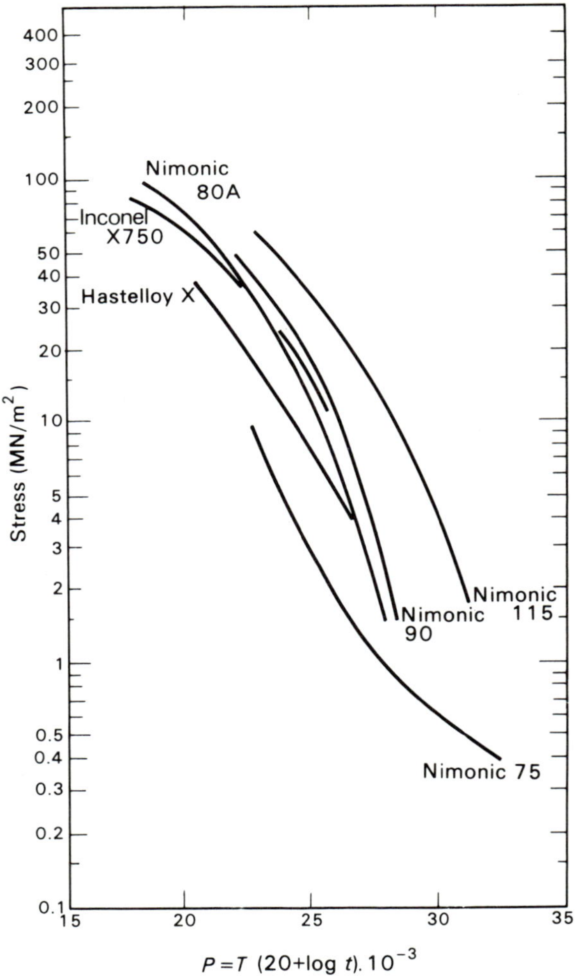

Fig. 23—Larson–Miller stress-rupture curves for alloys listed in Table 16 (T in kelvins, t in hours).

listed in Table 16 are shown in Fig. 23. The progressive increase in rupture strength of the alloys with increasing levels of matrix alloying elements and of titanium and aluminium contents is evident. Studies of the relationship between creep strength and composition and structure have shown that an important factor is the volume fraction of γ' phase in the alloy; this is shown in Fig.

24. The fraction of γ' phase is predominantly controlled by the titanium and aluminium contents, and hence progress in the development of improved alloys has very largely been achieved by increase of these contents. The difficulties to be overcome are mainly that the higher titanium and aluminium contents lead to entrapped oxide films and segregation, and that the greater high-temperature strength inevitably leads to increased difficulty in hot working the alloy to the required form. Improved melting techniques, such as electroslag refining and vacuum melting, have

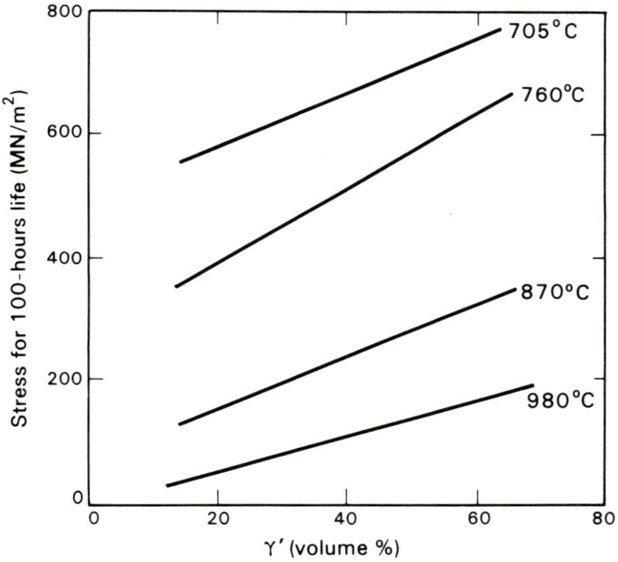

Fig. 24—Stress-rupture strength of nickel-base superalloys in relation to volume fraction of γ' precipitate.

contributed very largely to the progress made, and are dealt with in Chapter 5.

The short-time tensile properties of the alloys, so far as they affect the behaviour of the materials in service, are generally of significance only at temperatures up to about 500–600°C. At about this temperature and above, the time-dependent creep behaviour of the alloy becomes of major importance. However, the short-time properties at higher temperatures are relevant to manipulation of the alloys during hot working. The tensile properties of a typical alloy, Nimonic 80A, in the form of fully heat-treated bar,

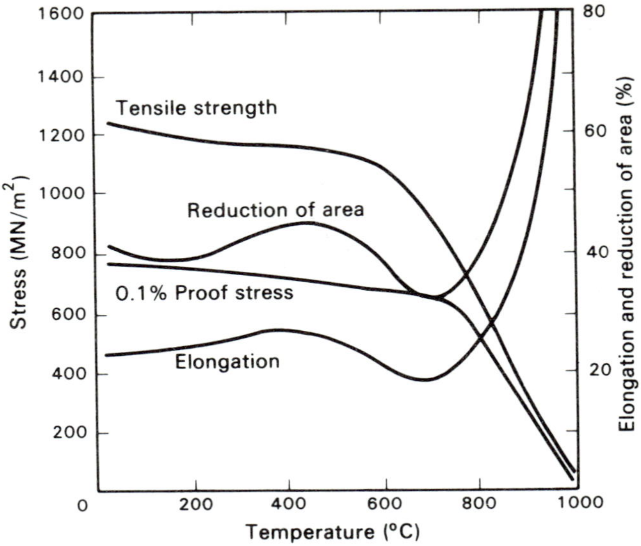

Fig. 25—Tensile properties of Nimonic alloy 80A.

are shown in Fig. 25. It is to be noted that the short-time properties fall only slowly with increasing temperature up to about 600°C, but then drop steeply. Alloys with increased creep strength at elevated temperatures generally show little increase in tensile strength at normal temperature but a progressive increase in the temperature at which their short-time tensile properties begin to

TABLE 18

Tensile properties of nickel-base superalloys—heat-treated bar.

Alloy	*20°C* 0.2% P.S. (MN/m²)	*20°C* Tensile strength (MN/m²)	*20°C* Elongation (%)	*800°C* 0.2% P.S. (MN/m²)	*800°C* Tensile strength (MN/m²)	*800°C* Elongation (%)
Nimonic 75	280	680	41	130	270	78
Hastelloy X	400	800	44	250	400	62
Nimonic 80A	770	1240	22	490	640	24
Nimonic 90	820	1250	27	600	690	10
Nimonic 115	850	1240	27	750	940	15
Inconel X 750	650	1130	24	360	440	15
Nimonic 901	900	1230	15	670	700	15

TABLE 19

Hardness of nickel-base superalloys.

Alloy	Hardness (HV) Annealed	Heat-treated
Nimonic 75	150–210	—
Nimonic 80A	200–250	290–370
Nimonic 90	230–300	340–400
Nimonic 115	370–440	370–440

fall. Hence at a temperature in the range 800–900°C, the alloys with increased creep resistance also show marked increase in short-time properties, as shown in Table 18.

The hardness of superalloys provides no guide to their high-temperature creep-resisting capabilities. Nevertheless, a know-

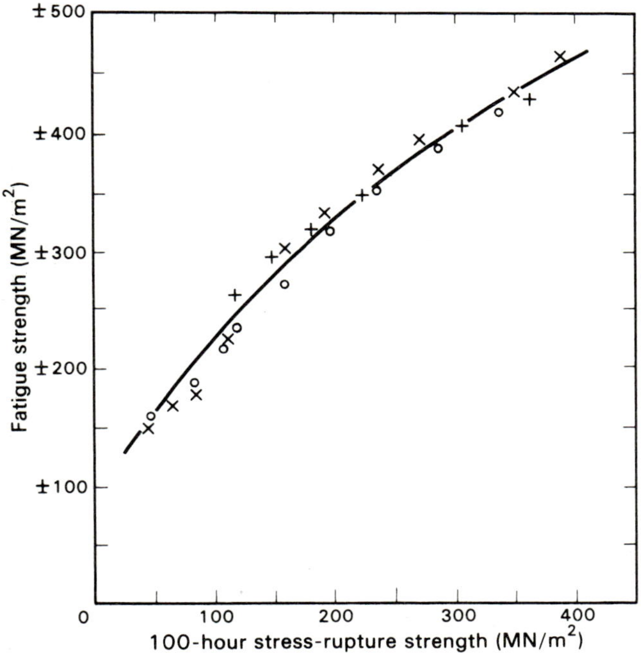

Fig. 26—Relationship between rotating-bend fatigue strength and stress-rupture strength of three different nickel-base superalloys.

ledge of the normal hardness of an alloy, and of its expected response to heat treatment, provides a ready means of checking for abnormalities and is, of course, a guide to one aspect of the alloy's machinability. The hardness values given in Table 19 may therefore be of interest.

Since articles under service at elevated temperatures may be subjected to fluctuating as well as steady stresses, the fatigue properties of the alloy at operating temperature may be as relevant as the creep properties. The fatigue properties of nickel-base superalloys have been measured under a variety of stressing conditions and temperatures. Within the temperature range over which creep phenomena occur, it has been shown that a direct relationship exists between the creep-rupture strength and the fatigue strength, for failure in the same time at the same temperature. Thus Fig. 26 shows the relationship between the rotating-bend fatigue strength and the stress-rupture strength for three alloys of the Nimonic series, which holds independently of the temperature or the time-to-failure. It applies for fatigue tests at 33–42 Hz over the temperature range 750–980°C, and for times up to about 1000 hours. The same principle applies for compression-tensile fatigue tests, although with a different detailed relationship.

Physical properties

Many of the physical properties of nickel-base superalloys are of importance in relation to their performance at high temperatures. In particular, their expansion coefficients, thermal conductivities and emissivities play a role in determining their response to thermal shock or thermal fatigue. These conditions arise with rapid or repeated changes of temperature caused by changes in temperature of the medium (liquid or gas) in which the alloy is immersed.

TABLE 20

Thermal expansion of nickel-base superalloys.

Temperature range (°C)	*Expansion coefficient*		($\times 10^{-6}$/°C)	
	Nimonic 75	*Nimonic 80A*	*Nimonic 90*	*Nimonic 115*
20–100	11.0	11.1	11.5	11.1
20–200	12.7	12.5	12.4	11.8
20–400	13.9	13.7	13.5	12.6
20–600	15.0	14.8	14.6	13.7
20–800	16.5	16.3	16.5	15.2
20–1000	18.2	18.8	19.2	17.9

In particular, they arise in the operation of gas turbines due to the change in temperature of the gas passing through the turbine. Thermal shock or fatigue properties are normally assessed by tests simulating the appropriate service conditions. The thermal expansion behaviour of some of the Nimonic alloys is given in Table 20. The elastic moduli of the alloys are little affected by the changes in composition between the alloys of the series, and typical values, determined by a dynamic method which avoids the difficulties introduced by creep under a steady load, are as given below:

Temperature (°*C*)	20	100	200	400	600	800	1000
Young's modulus (GN/m^2)	221	216	210	197	183	166	141

The thermal conductivities of the alloys are all low in comparison with those of pure metals, and lie in the range 0.10–0.13 W/m K at normal temperature, rising to 0.22–0.26 W/mK at 800°C. The total emissivity of the oxidised alloys is high, reaching values approaching 1.0 when measured at temperatures exceeding 700°C after oxidation at about 1200°C.

Cast alloys

There are two main reasons for the increasing preference for the use of cast rather than wrought nickel-base superalloys for articles required to operate at the very highest possible temperature, e.g. the rotor blades of aircraft gas-turbines. The first is that, to take advantage of the higher stress-rupture properties developed with increasing titanium and aluminium contents, the temperature at which the alloy must be hot worked increases until it becomes so close to the solidus that there is insufficient latitude for adequate control. The second is that a forging has inherently lower creep properties than a casting of the same composition, by virtue of the fact that the forging deformation has introduced defects into the crystal lattice which initiate creep. Special casting alloys have therefore been developed with higher levels of titanium and aluminium (4–6 per cent of each) which cannot be hot worked and which need to be melted and cast in vacuum. Typical compositions were included in Table 17. The stress-rupture properties of such alloys may be such as to permit operation at temperatures around 1100°C. Figure 27 shows, for example, the Larson–Miller plots, on the same scale as those for the wrought superalloys shown in Fig. 23, of the stress-rupture properties of the four casting alloys listed in Table 17. Casting alloys are normally used in the as-cast condition, since further heat treatment can do little to improve the creep properties. The properties do, however, depend to a limited extent on the grain size and on the size of the test bar used. The coarser the grain size, the lower the creep-rate, but there is a reduction in the creep extension before fracture which leads to uncertainty in the effect on stress-rupture life. The lives are more consistent if the grain size is such that there are

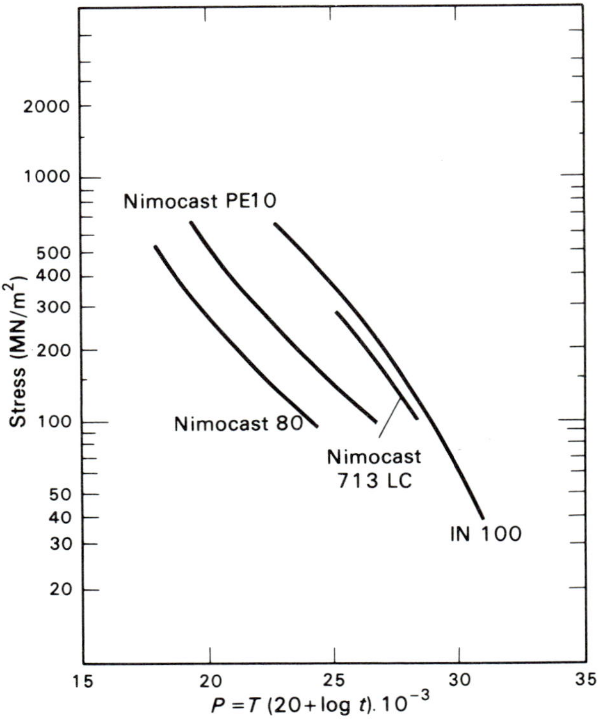

Fig. 27—Larson–Miller stress-rupture curves for cast nickel-base superalloys (T in kelvins, t in hours).

many grains in the cross-section of the test piece or component, or alternatively if the test piece consists of a single crystal—clearly the grain boundary is a source of fracture weakness. Improved service performance is therefore being sought by the introduction of special casting techniques to give single-crystal castings, columnar growth of crystals in the direction of major service stress, or controlled grain-size castings.

CORROSION

The high oxidation resistance of the simple nickel–chromium binary alloys from which the superalloys are developed is generally maintained in the more complex materials, but other corroding agents may occur in high-temperature service. Particularly important are sulphur-containing compounds, arising from fuels or from the combustion air, and vanadium, an impurity in some residual mineral oils. The mechanisms of attack by such compounds were discussed on pp. 45–47.

So far as oxidation in clean air is concerned, the hardening elements, particularly the titanium and aluminium, have a relatively small effect, although the elements in solid solution, notably molybdenum and cobalt, may reduce the scaling resistance. Table 21 compares the oxidation loss for three typical alloys between 800 and 1000°C.

TABLE 21

Oxidation loss for three typical alloys.

Alloy	*Weight loss* (g/m^2) *after 100 h at temperature and descaling*			
	800°C	*900°C*	*950°C*	*1000°C*
Nimonic 75	5.5	11.8	40.0	66.6
Nimonic 80A	6.4	26.2	39.6	59.6
Nimonic 90	4.6	25.2	55.0	104.0

Sulphur corrosion is a recurrent problem with nickel-base superalloys, leading to attack of the type described earlier (p. 45). It is normally kept in check by ensuring an adequate chromium content in the alloy (rarely less than 15 per cent), avoiding reducing conditions in the atmosphere and, if necessary, by protection of the superalloys with metallic (aluminium-silicon) or refractory-oxide coatings. Attack by vanadium can be controlled to some extent by similar measures, and by the use of inhibitors added to the combustion system, but it is normally avoided by the selection of low-vanadium oils for use in high-temperature plant.

NICKEL–MOLYBDENUM ALLOYS

Nickel readily takes molybdenum into solid solution, the solubility being about 20 weight per cent at normal temperature and rising

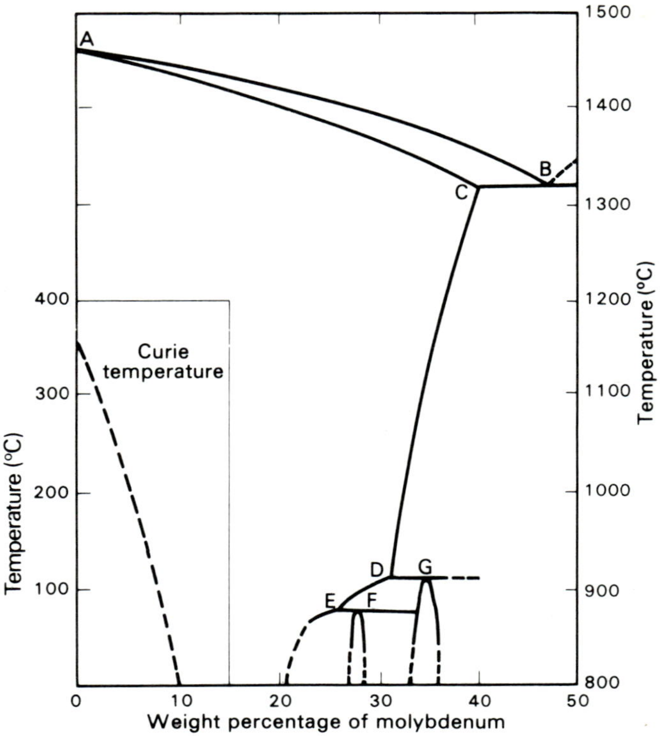

Point	A	B	C	D	E	F	G
°C	1453	1318	1318	910	875	875	910
%Mo	0	47	40	32	26	28	35

Fig. 28—Nickel–molybdenum equilibrium diagram.

to almost 40 per cent at 1315°C (a eutectic is formed melting at the higher temperature). The nickel-rich end of the constitutional diagram is shown in Fig. 28. The principal effect of molybdenum additions is to improve the corrosion resistance to a wide range of media, and to strengthen the metal, particularly at elevated temperatures. The latter influence can be detrimental to the workability of the alloys so that on technical grounds, as well as economic ones, the level of molybdenum content in commercial alloys is generally restricted to a figure well below the solubility limit.

The simplest nickel-base alloy containing molybdenum is essentially a binary alloy composed of about 28 per cent molybdenum, although the iron content may be up to about 5 per cent. This alloy, termed Hastelloy B, is particularly notable for its high resistance to corrosion by hydrochloric acid in all concentrations up to the boiling point, and to other acid solutions other than oxidising ones. Thus in sulphuric acid the corrosion rates in all concentrations of acid at normal temperature do not exceed 25 μm/year; in boiling acid the rates are low (less than 100 μm/year) at all concentrations up to 60 weight per cent, but then increase rapidly. Under all conditions the corrosion rates are increased by aeration or by the presence of oxidising salts. Similarly, Hastelloy B is the most resistant of nickel-base materials

TABLE 22

Compositions of molybdenum-containing nickel-base alloys.

Alloy	*Mo*	*Cr*	*Fe*	*W*	*Cu*	*Ni*
Hastelloy B	28	—	5	—	—	bal.
Hastelloy W	25	5	5	—	—	bal.
Hastelloy C	17	15	5	4	—	bal.
Hastelloy N	17	7	5	—	—	bal.
Hastelloy X	9	22	18	—	—	bal.
Ilium G	6.5	22.5	6.5	—	6.5	bal.

to phosphoric acid, the corrosion rate being below 50 μm/year for concentrations up to 85 per cent at temperatures up to 66°C, and below 100 μm/year for concentrations up to 50 per cent at temperatures up to the boiling point. Hastelloy B also retains its strength well at elevated temperatures, although it is not so resistant to oxidation as the chromium-containing alloys. On account of its good strength retention it was used in the U.S.A. for the blades of aircraft-exhaust superchargers, and indeed was used for turbine blades in one of the earliest American jet engines until higher-strength superalloys were developed.

Other molybdenum-containing alloys also have additions of chromium to improve the resistance to corrosion, particularly to oxidation. A range of alloys with varying molybdenum/chromium ratios is available, typical alloys being listed in Table 22. In general, the higher-molybdenum alloys have the best wet corrosion resistance, except to oxidising media, while the high-chromium alloys have the highest resistance to oxidation, either wet or dry. Hastelloy C has good resistance to a wide range of corrosive media including oxidising solutions, especially those containing chlorides, although its resistance to pure unaerated sulphuric acid at temperatures above ambient is not so good as that of Hastelloy B; at normal temperature, however, its corrosion rate in sulphuric acid of all concentrations does not exceed 10 μm/year. In phosphoric acid it has excellent resistance at temperatures up to 66°C and reasonable resistance up to the boiling point. Hastelloy N has intermediate levels of both chromium and molybdenum and was developed to have optimum properties in use with the molten fluoride salts in nuclear plants operating at temperatures in the range 700–900°C. Hastelloy X is primarily a high-temperature alloy and has already been referred to on p. 52.

Additions of copper up to about 7 per cent are sometimes made to nickel–molybdenum–chromium alloys containing 20–25 per cent chromium and 5–10 per cent molybdenum. These form the Ilium alloys, mainly used in the cast form, although at the lower limits of alloying additions they can be wrought. Their main applications are as materials resistant to corrosion by strong mineral acids, including oxidising ones, and to sea-water.

NICKEL–IRON ALLOYS

By far the greatest commercial use of nickel in association with iron is as an alloying element in steels and cast iron. Nickel is used in low-alloy steels at levels at 1–5 per cent, either alone or in combination with other elements. Its main effect is to retard the martensitic transformation and thus increase the hardenability of the steel. It also increases the impact strength, particularly at sub-zero temperatures. Nickel is an essential constituent to the extent of 7 per cent or more in austenitic stainless steels, typified by the most common one containing 8 per cent nickel and 18 per cent chromium. Similarly in the cast-iron field, additions of nickel up to about 3 per cent are made to cast irons to give increased strength and toughness, and at much higher levels (in the region of 20 per cent) to austenitic irons. It is also a common vehicle for the introduction of magnesium into cast iron to develop spheroidal graphite structures. A nickel–magnesium alloy containing about 25 per cent magnesium enables the magnesium to be introduced into the molten iron without a violent reaction. The residual nickel confers on the resultant iron the benefits of increased strength and toughness. All these applications, however, fall within the field of ferrous metallurgy and are not therefore dealt with in any detail here.

The constitutional diagram of the nickel–iron system is shown in Fig. 29, from which it can be seen that at temperatures between about 900 and 1400°C the two metals form a continuous series of face-centred-cubic solid solutions. The A_3 transformation tempera-

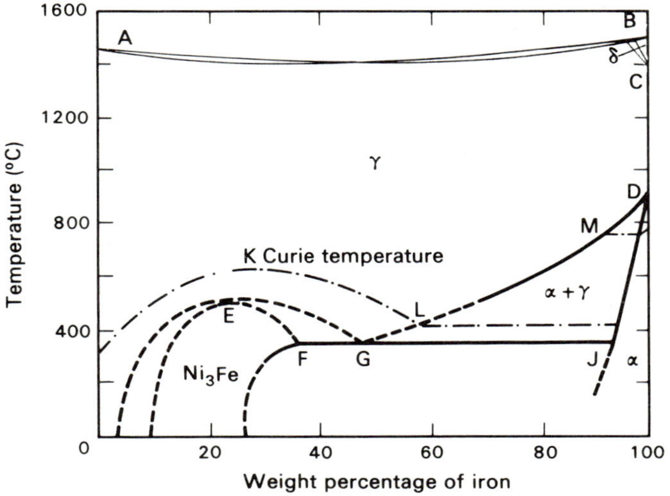

Point	A	B	C	D	E	F	G	J	K	L	M
°C	1453	1535	1390	910	503	345	345	345	612	403	765
%Fe	0	100	100	100	24	36	47	93.2	~30	–	–

Fig. 29—Nickel–iron equilibrium diagram.

ture of iron (from face-centred-cubic austenite to body-centred-cubic ferrite) is depressed by additions of nickel, and the transformation also becomes very sluggish so that there is a wide difference in the transformation temperatures on heating and on cooling. The difficulty of obtaining equilibrium conditions at temperatures below about 400°C means that in practice all binary alloys containing more than about 30 per cent nickel retain their austenitic structure at normal temperature. Both ferritic and austenitic phases are ferromagnetic at normal temperature, the Curie point being about 770°C for pure-iron ferrite and falling only slightly with increasing nickel content. However, that for austenite rises from 358°C for pure nickel to a peak at just over 600°C at about 30 per cent iron; with further reduction in nickel content the Curie point falls until it is close to normal temperature at 30 per cent nickel. Indeed nickel–iron alloys of this composition may be non-magnetic at normal temperature, depending on the minor contents of other elements and on the thermal history. They are therefore of value as temperature-compensating alloys in magnetic devices and will be referred to again later (p. 68).

Although the magnetic response of nickel is considerably less than that of iron, the magnetic saturations at normal temperature being about 0.6 T (6000 gauss) and 2.0 T (20 000 gauss) respectively, the alloys of the binary nickel–iron system have valuable magnetic characteristics which have led them to be used for a number of practical applications, and also to be the basis of more complex alloys with special physical properties[5]. The alloys used particularly for their magnetic properties may be divided into the soft, high-permeability materials and the hard, permanent magnet materials, while the main groups of other alloys are those with controlled expansion properties and those with constant elastic moduli.

Soft magnetic alloys

Soft magnetic materials are those which are readily magnetised to saturation in a weak magnetic field and which also readily lose their magnetism on removal of the field. The particular characteristics which make an alloy of merit for magnetic applications vary widely, depending on the details of the application, and embrace the initial permeability at low field strengths, the incremental permeability at specific levels of field strength, the effects of frequency of fluctuating fields, and hysteresis losses, etc. (for a full appreciation of the technology of such materials, reference should be made to specialised textbooks[5,6]). However, an understanding of the importance of nickel–iron alloys in this field can be obtained by considering the data shown in Figs. 30 and 31. Figure 30 shows the initial permeability of binary alloys in three conditions of heat treatment. The exceptionally high value at about 78 per cent nickel is of particular interest—this forms the basis of a group of alloys termed "Permalloy". Figure 31 shows the magnetic induction at a high field strength giving effective saturation, and the high value at about 50 per cent nickel, coupled with reasonable permeability, makes this alloy also of commercial interest. The properties of both groups of alloys depend critically

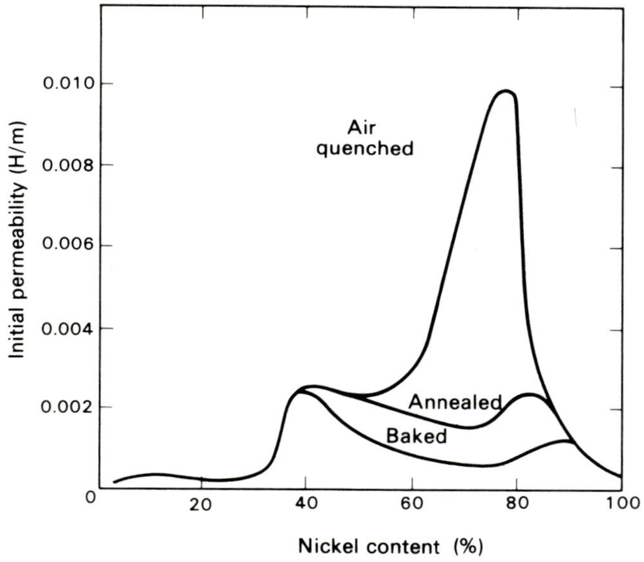

Fig. 30—Initial permeability of nickel–iron alloys.

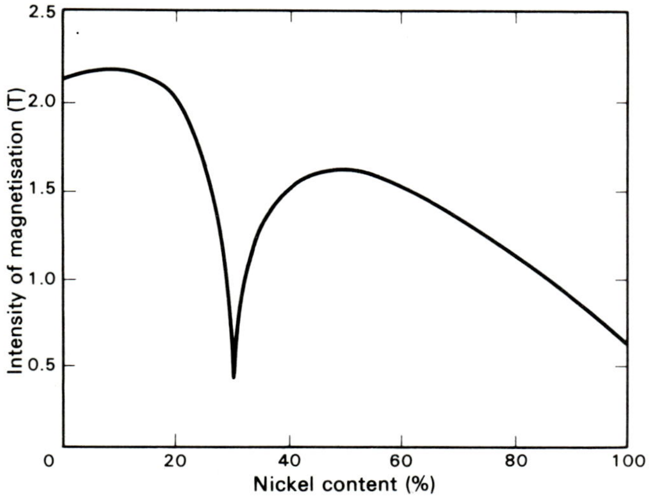

Fig. 31—Magnetic induction of nickel–iron alloys with a magnetising field of 32 kA/m.

upon purity and on heat treatment, and also on the extent of preferred orientation developed during cold working to strip or laminations, the forms in which they are normally used. Indeed, special versions of the alloys are produced by developing grain-oriented structures during cold working and heat treatment; in some cases heat treatment in a magnetic field is adopted in order to give a particular orientation to the magnetic domains. The need for precise control of composition, and particularly the elimination of undesirable impurities such as silicon, has led to the adoption of powder metallurgical techniques for the production of many of the alloys of this type, since suitable high-purity powders of both nickel and iron are available.

Typical properties of the main types of soft, magnetic nickel–iron alloy in comparison with those of pure iron and nickel are given in Table 23.

TABLE 23

Typical magnetic data of high-permeability alloys.

Composition (%)	*Initial permeability* (10^{-3} *H/m*)	*Maximum permeability* (10^{-3} *H/m*)	*Saturation induction* (*T*)	*Curie temperature* (°*C*)	*Resistivity* ($10^{-8}\,\Omega m$)
99·98 Fe	3.8	38	2·05	770	12·0
22 Fe, 78 Ni	19–50	60–190	0·65–0·8	300–450	55–60
33 Fe, 65 Ni, 2 Mo (domain oriented)	0·5–1·3	2·5–500	1·40	590	26
50 Fe, 50 Ni	2·5–5·0	25–50	1.50–1.60	350–500	45–55
99 Ni	0.25–0.38	1.9–3.2	0.6	358	6.8

Many electrical measuring and control devices need the incorporation of a mechanism to compensate for the effect of changing temperature, which otherwise would disturb their accuracy. This may be effected by a magnetic shunt made of an alloy whose permeability is particularly sensitive to changes of temperature within the range to be covered. Alloys operating just below their Curie point are most suitable since the permeability is then most sensitive to temperature. For temperature compensation near to normal temperatures, alloys with Curie points in the range 60–100°C are used, and nickel–iron alloys with about 30 per cent nickel form the basis of the most commonly used commercial alloys. Minor additions of other elements and special working and heat-treatment procedures may also be involved. Nickel–copper alloys containing about 70 per cent nickel (*see* p. 41) are also used in this connection.

Permanent magnet alloys

Until about forty years ago permanent magnets were made of alloy steels in which the martensitic transformation provided the disturbance of the basic ferromagnetic structure, which is now known to be necessary to develop a high coercive force. Improved magnetic properties had been developed empirically, starting from plain carbon steels and progressing through tungsten, chromium and cobalt steels. In 1932 Mishima announced the discovery of a permanent magnet alloy of approximate composition 25 per cent nickel, 12 per cent aluminium, balance iron, with much improved magnetic properties, particularly the coercive force. Developments of this basic alloy now form by far the most important group of permanent magnet materials.

Structurally, alloys of this type consist of a body-centred-cubic matrix in which a dispersed precipitate of a second body-centred-cubic phase is developed by heat treatment. The second phase has an ordered structure based on Fe_2NiAl and a lattice parameter only slightly different to that of the matrix. The high coercive force of the alloy is associated with the correct dispersion of the precipitate. Progressive improvement in the properties of alloys of this type has been obtained by additions of other elements, particularly cobalt, by heat treatment in a magnetic field, and by control of the grain structure of the material by the growth of columnar crystals.

The alloys have relatively poor mechanical properties, being coarse-grained, brittle and hard, so that they cannot be hot worked. They are normally used as castings but can also be formed to the required shape by powder metallurgical techniques. The magnetic properties of a permanent magnet material are characterised by the parameters of the demagnetisation curve illustrated in Fig. 32. The remanence, B_r, is the magnetic flux remaining after removal of the magnetising field, and the coercive force, H_c, is the reverse field required to reduce the flux to zero. In practice, a permanent magnet operates at some point between these two limits as a result of the demagnetising action of the magnetic circuit in which the magnet is serving. For the most efficient use of the magnet material, it is desirable that it should be designed to work at the point at which the product $B \times H$ is a maximum, i.e. at the point marked A. The quantity BH_{max} represents the stored magnetic energy in unit volume and is a useful criterion for the efficiency of a permanent magnet material. Table 24 gives the magnetic characteristics of typical magnet alloys.

To develop the optimum properties of these permanent magnet

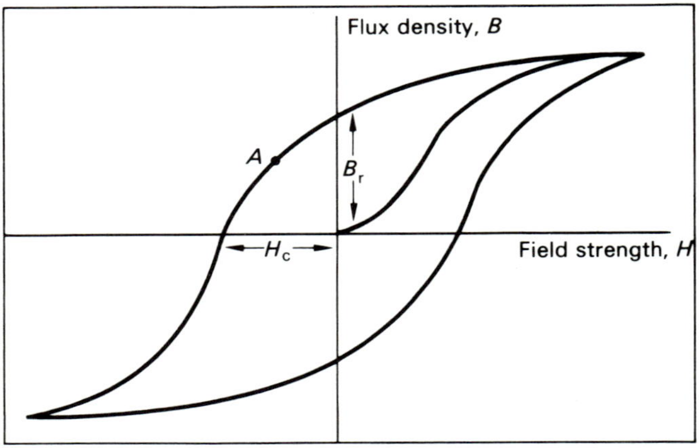

Fig. 32—Hysteresis loop of permanent magnet alloy. At *A* the product *BH* is a maximum.

TABLE 24

Compositions and properties of permanent magnet alloys.

Type	*Composition (bal. Fe)*				*Remanence*	*Coercive force*	BH_{max}
	Ni	*Co*	*Al*	*others*	*(T)*	*(10⁴ A/m)*	*(10⁴ J/m³)*
Chromium steel	—	9	—	3Cr, 1C	0.95	0.48	0.24
Tungsten steel	—	7	—	6W, 1C	1.05	0.52	0.24
Cobalt steel	—	35	—	6Cr, 5W, 1C	0.86	1.80	0.70
Alni	25	—	12	5Cu, 1Ti	0.60	4.00	1.00
Alnico	20	12	10	5Cu, 1Ti	0.75	5.00	1.40
Alcomax	16	22	9	3Cu, 1Ti	1.20	6.00	3.80
Hycomax	15	35	8	3Cu, 3Ti	1.00	10.00	6.00

alloys, it is essential that they are correctly heat-treated. The details of the treatment depend critically on the exact composition. In general, however, all treatments involve an initial solution treatment at a temperature in the range 1000–1200°C, followed by cooling, sometimes at a controlled rate to nucleate precipitation.

Precipitation is then completed by reheating to 550–650°C for a period of a few hours. The effects of heat treatment in a magnetic field are to produce anisotropic magnetic properties—the properties in the direction of the field being improved at the expense of those in the transverse direction. The stage of treatment during

Permanent Magnet Association

Fig. 33—Equiaxed and columnar structures of a permanent magnet alloy. *Top:* equiaxed structure with isotropic magnetic properties; *bottom:* columnar structure with improved properties along crystal growth direction.

which the magnetic field is effective is that of the nucleation and growth of the precipitated phase. Hence the field is normally applied during cooling from the solution treatment. It may also be advantageous to reheat to the precipitation temperature in a magnetic field. Columnar crystal magnets are made by adopting special compositions and solidification conditions which ensure the growth of long columnar crystals oriented with a preferred magnetic axis along the direction of growth of the crystals. Again the resultant magnetic properties are anisotropic. The grain structure of such alloys in comparison with that of a normal isotropic magnet is shown in Fig. 33.

Controlled-expansion alloys

The coefficient of linear thermal expansion of binary nickel–iron alloys shows a marked anomaly in the region of the boundary between the ferritic and austenitic structures. Figure 34 shows the expansion coefficient at 20°C plotted against nickel content—the alloy with 36 per cent nickel has a minimum coefficient of about 1×10^{-6} K^{-1}, and is widely known as "Invar". The low coefficient only applies up to moderate temperatures, as illustrated in Fig. 35, and for wider ranges of temperature higher-nickel alloys have lower average coefficients. The 36 per cent nickel alloy, however, maintains its low-expansion characteristics down to absolute zero and is, therefore, of particular interest in cryogenic engineering. The anomalous expansion of these alloys is associated with the magnetic properties, the low expansion being a balance between normal thermal expansion and contraction due to magnetostric-

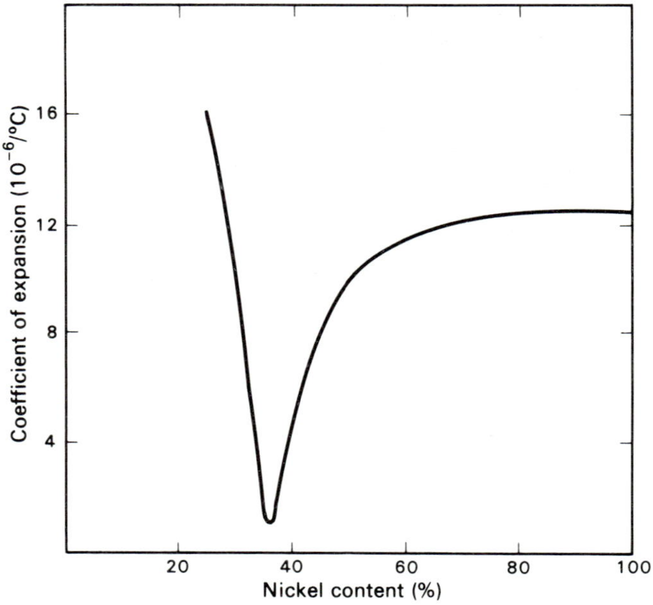

Fig. 34—Expansion coefficient at 20°C of nickel–iron binary alloys.

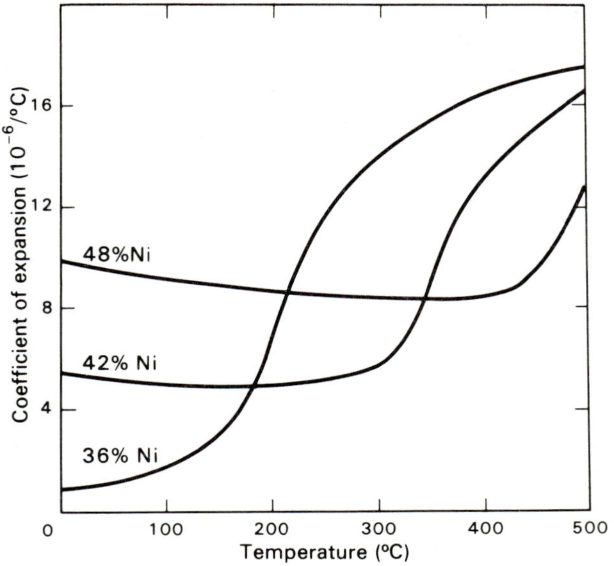

Fig. 35—Expansion coefficients of nickel–iron binary alloys at different temperatures.

tion. At temperatures above the Curie point (*see* Fig. 29) the alloys are, of course, non-magnetic and the expansion coefficient reverts to normal.

The precise value of the expansion coefficient, apart from being critically dependent on the nickel–iron ratio, is also affected by the presence of other elements, which generally increase the coefficient; by annealing, which increases the coefficient; and by cold work, which can decrease the coefficient—indeed the 36 per cent nickel alloy can be cold worked sufficiently to give a negative expansion coefficient. Additions of cobalt are of particular interest, since they reduce the expansion coefficient and at the same time increase the Curie temperature, thus widening the temperature range over which the low coefficients apply. The most important alloy of the ternary series contains about 17 per cent cobalt and 29 per cent nickel, and is used in sealing to hard glasses of the borosilicate type. Alloys of the nickel–iron–chromium system are also used for making glass-to-metal seals, the advantage in this case being that the oxide film derived from the chromium aids in forming a bond between the glass and the metal. The expansion coefficient is slightly increased in comparison with that of the corresponding chromium-free alloy, but the alloy is suitable for use with the softer, higher-expansion glasses. Typical compositions of a selection of controlled-expansion alloys with the coefficients for appropriate temperature ranges are given in Table 25.

TABLE 25

Compositions and properties of controlled-expansion alloys.

Composition (bal. Fe)			*Expansion coefficient (10^{-6} per °C)*					
Ni	*Co*	*Cr*	*20–100*	*20–200*	*20–300*	*20–400*	*20–500*	*20–600*
36	—	—	1.2	2.3	4.8	7.5	9.1	10.0
42	—	—	4.7	4.4	4.6	5.8	7.6	8.8
50	—	—	8.7	9.4	9.3	9.1	9.7	10.8
29	17	—	5.9	5.2	5.1	5.1	6.2	7.8
48	—	5	9.2	9.3	9.3	10.1	11.3	12.1

It has recently been found possible to confer precipitation hardening characteristics on some of the controlled-expansion alloys by the addition of small percentages of titanium. The hardening is due to the precipitation of γ' phase, akin to that occurring in nickel-base superalloys. The tensile strength of alloys of this type, exemplified by Incoloy 903, can be increased to 700 MN/m^2, sufficiently high to make them useful for highly-stressed engineering parts, e.g. shafts in aircraft gas-turbines.

Constant-modulus alloys

With most alloys, the elasticity moduli fall with increasing temperature. With nickel–iron alloys, however, associated with the low expansion coefficients, it is found that there are anomalous thermoelastic coefficients as illustrated in Fig. 36. For alloys with between 27 and 43 per cent nickel there is a positive coefficient, so that the alloy becomes stiffer with increase in temperature, the maximum coefficient occurring at about 36 per cent nickel. Additions of other elements make the nickel–iron ratio less critical and increase the strength of the alloy, so that commercial alloys exploiting this characteristic are usually ternary or more complex alloys based on the nickel–iron–chromium or nickel–iron–molybdenum systems. The effect of chromium additions is shown by the second curve in Fig. 36. These alloys can also be further strengthened by the inclusion of small amounts of titanium and aluminium, and a typical alloy of this type is Ni-Span C 902, containing 42 per cent

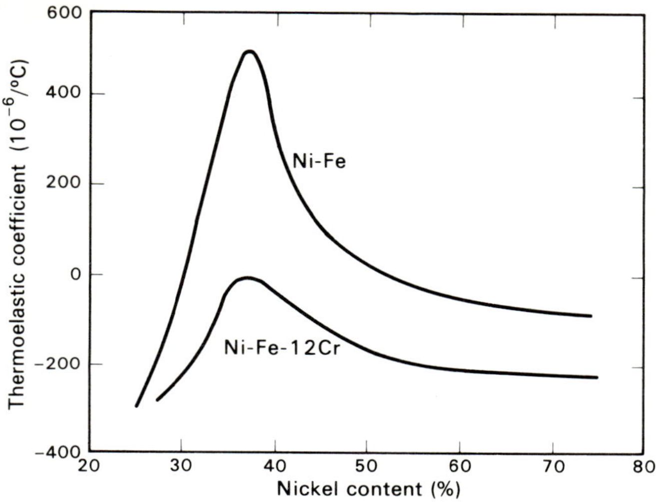

Fig. 36—Temperature coefficients of Young's modulus for nickel–iron and nickel–iron–chromium alloys

nickel, 5.2 per cent chromium, 2.4 per cent titanium and 0.6 per cent aluminium. After appropriate heat treatment this alloy has the following properties:

hardness (HV)	420
0.2% proof stress (MN/m^2)	1200
tensile strength (MN/m^2)	1450
elongation (per cent)	5
expansion coefficient (20–100°C, K^{-1})	6.0×10^{-6}
Young's modulus (GN/m^2)	185
thermoelastic coefficient ($10^{-6} K^{-1}$)	0–50

Constant-modulus alloys generally have low Curie points, and thus considerable choice in magnetic properties can be exercised. Non-magnetic alloys can be selected for such applications as watch hair-springs, or alloys with a high magnetic response can be chosen for magnetically-operated devices—there is a wide variety of applications in the field of measuring devices and oscillating mechanisms.

References

1. Hansen, M., *The Constitution of Binary Alloys*, 2nd edn., McGraw Hill, New York, 1958; 1st suppl., Elliott, R. P., 1965; 2nd suppl., Shunk, F. A., 1969.

2. Betteridge, W., and Heslop, J., *The Nimonic Alloys*, 2nd edn., Arnold, 1975.
3. Sims, C. T., and Hagel, W. C., *The Superalloys*, Wiley, New York, 1972.
4. Larson, F. R., and Miller, J., *Transactions of the American Society of Mechanical Engineers*, vol. 74, p. 765, 1952.
5. Bozorth, R. M., *Ferromagnetism*, Van Nostrand, New York, 1951.
6. Hoselitz, K., *Ferromagnetic Properties of Metals and Alloys*, Clarendon Press, 1952.

CHAPTER FOUR

Metallography

The metallography of nickel and nickel alloys presents few unique problems, although the wide range of alloys to be considered does make it necessary to have a full range of techniques available if the best results are to be obtained. In all cases, the normal precautions recommended for all metallographic studies must be carefully observed in order to ensure that the true structure of the material is revealed. In essence, these consist of ensuring that, at each successive stage of the polishing operation, the disturbance of the structure caused by the cold-working effects of the previous stage are completely removed, and that all surface contamination is removed before etching. Given these essentials, there is usually no difficulty in revealing the structure of nickel alloys, although, because of their relatively high corrosion resistance, it is frequently necessary to use vigorous etching reagents.

PREPARATION OF SPECIMENS

Abrasive polishing of macro- and micro-specimens may be carried out in the normal manner using graded silicon-carbide papers, but the process can be considerably shortened by using 3-μm diamond paste on cloth-covered laps for the later stages. Final polishing may be effected with a water suspension of fine gamma alumina on a rotating cloth-covered wheel, or by 1-μm diamond paste.

As an alternative to abrasive polishing, electrolytic polishing is frequently advantageous, and several different solutions and procedures have been recommended (*see* Table 26). The specimen is made the anode, and a platinum or stainless-steel cathode is used. The conditions of voltage, current density and time quoted are those found to be satisfactory, but in general, experimental variation of these must be made until the best results on any particular material are obtained.

TABLE 26

Electropolishing of nickel-base alloys.

Alloy	*Electrolyte* (*vol %*)		*Temperature* (°*C*)	*Voltage*	*Current density* (A/cm^2)	*Time*
Nickel, Ni–Cu, Ni–Fe, Ni–Fe–Cr	Phosphoric acid	37				
	Glycerol	56	—	—	1–3	1–4 min
	Water	7				
Nickel, Ni–Cu, Ni–Cr, Ni–Fe–Cr	Nitric acid	33	20–30	40–50	1–2.5	25 s
	Methanol	67				
Ni–Cr, Inconel alloys	Phosphoric acid	25				
	Nitric acid	25	—	—	15–20	5–10 s
	Water	50				
Nimonic alloys	Glacial acetic acid	89				
	Perchloric acid (Δ1.59)	8	25	15	0.1	3–4 min
	Water	3				
	Perchloric acid (Δ1.20)	20				
	Ethyl alcohol	70	—	40	1	15–20 s
	Glycerol	10				

ETCHING

For macro-etching of nickel alloys to reveal grain structure, vigorous reagents based on strong mineral acids are necessary. Table 27 lists some of those found suitable for different classes of alloys.

For micro-etching it is also necessary to use vigorous reagents, and a wide range of these has been given in the literature[1]. Electrolytic etching is generally more controlled in action and to be preferred. Selected reagents and conditions of use, both for immersion and electrolytic etching, are given in Table 28. The exact conditions for satisfactory etching are critically dependent on many factors which are not always precisely controlled, and

TABLE 27

Macro-etching reagents for nickel-base alloys.

Alloy	*Solution* (*vol* %)		*Conditions*
Nickel	Nitric acid Acetic acid	50 50	—
Ni–Cu	Nitric acid Water	50 50	—
Ni–Cr, Ni–Fe–Cr	Hydrochloric acid Hydrogen peroxide (30%) Water	40 40 20	Use fresh. Add peroxide as necessary
	Nitric acid Hydrochloric acid	25 75	—
Nimonic alloys	Hydrochloric acid Water	50 50	Use boiling
	Hydrochloric acid Chromic acid (10%)	50 50	Use boiling
	Hydrochloric acid Water Ferric chloride	40 60 200 g/l	For grain size

hence it is normal practice to determine the best conditions by experiment. Ambient temperature, age of etching solution, period of exposure of sample to atmosphere, aeration of solutions, etc., are all factors which may influence the effectiveness of the operation—the skill, experience and judgment of the operator play a large part in obtaining satisfactory results.

STRUCTURES OF NICKEL-CONTAINING MATERIALS

A great many of the nickel-based alloys referred to in Chapter 3 are essentially single-phase structures which do not, therefore, present interesting metallographic characteristics other than grain size and morphology. Nevertheless, the monitoring of the struc-

TABLE 28

Micro-etching reagents for nickel-base alloys.

A. Immersion etching.

Alloy		*Solution (vol parts)*		*Conditions*
Nickel, Ni–Cu	(*i*)	Sodium cyanide	1	
		Ammonium thiosulphate	1	
		Water	2	
	(*ii*)	Nitric acid	1	
		Hydrochloric acid	3	
		Glycerol	2	
	(*iii*)	Nitric acid	1	
		Acetic acid	1	
		Acetone	1	
	(*iv*)	Nitric acid	2	
		Acetic acid	1	
		Water	1	
	(*v*)	Nitric acid	20	1–3 s for grain size
		Hydrofluoric acid	3	
Nickel-silver		*See* (*iii*) above		
		See (*iv*) above		
	(*vi*)	Potassium cyanide	5	For low-zinc alloys
		Water	95	
		Hydrogen peroxide	*few drops*	
	(*vii*)	Ammonium hydroxide	85	For high-zinc alloys
		Hydrogen peroxide	15	
Ni–Cr		*See* (*ii*) above		
		See (*v*) above		
	(*viii*)	Hydrochloric acid	3	
		Nitric acid	·1	
		Cupric chloride	*saturate*	
Nimonic alloys	(*ix*)	Cupric chloride	80 g	
		Ferric chloride (hydrated)	60 g	
		Hydrochloric acid	85 ml	
		Water	100 ml	

TABLE 28 (*cont.*)

B. Electrolytic etching.

Alloy		*Electrolyte* (*vol %*)		*Voltage*	*Current density* (*A/dm*²)	*Time*
Nickel, Ni–Cu	(*a*)	Phosphoric acid Water	20 80	+20	—	10–15 s
	(*b*)	Sulphuric acid Hydrogen peroxide Water	22 12 66	—	—	—
	(*c*)	Nitric acid Acetic acid Water	10 5 85	+1.5	—	—
	(*d*)	Sulphuric acid Water	5 95	—	—	—
Ni–Cr		*See* (*d*) above		—	—	—
	(*e*)	Hydrochloric acid Water	10 90	+10	10	8–10 s followed by (*f*)
	(*f*)	Sodium hydroxide Water	10 90	−10	24	5–10 s following (*e*)
Nimonic alloys	(*g*)	Hydrofluoric acid Glycerol Water	5 10 85	+2–4	0.5	5–15 s
	(*h*)	Nitric acid Methanol	20 80	+30	—	15–30 s
	(*j*)	Hydrochloric acid Water	50 50	—	15–60	5 min for macro-etch

tures of such materials is important in order to ensure that a degree of consistency in properties is maintained. Comments on the structures of these materials are presented in the following sections which deal with each of the main groups of alloys dealt with in Chapter 3.

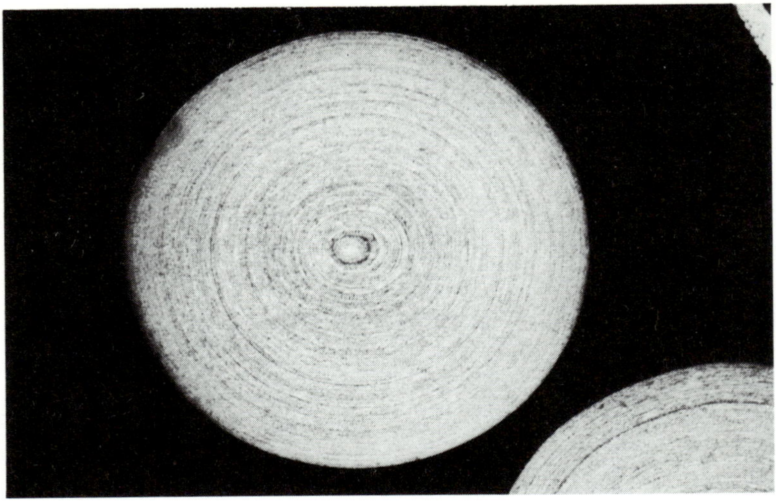

(*a*)

(*b*)

[*Inco Europe Ltd.*

Fig. 37—Microstructure of primary nickel: (*a*) carbonyl pellet (×10); (*b*) electrolytic cathode (×35).

PURE NICKEL

Before dealing with the more familiar forms of wrought nickel, it is interesting to note the structure of the two common forms of primary nickel—electrolytic cathode and carbonyl pellets—shown in Fig. 37. In each case the growth pattern is delineated by lines perpendicular to the growth direction formed by fluctuations in the deposition conditions.

Carbonyl nickel powders do not show growth rings since each grain is formed in a single step of decomposition and, unlike the pellets, is not built up by successive steps. Depending on the exact conditions of decomposition, the powder grains vary from relatively smooth, spherical particles to spiky particles, or to agglo-

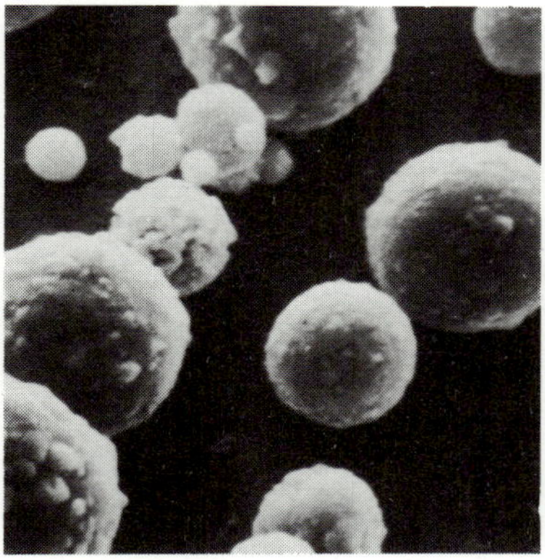

(*a*)

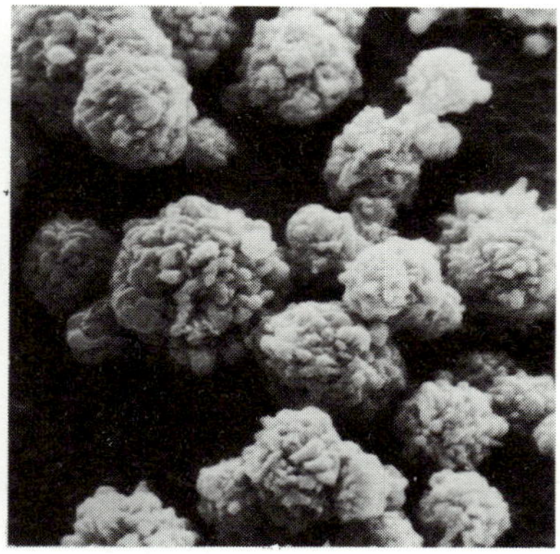

(*b*) [*Inco Europe Ltd.*

(*c*)

Fig. 38—Scanning electron micrographs of carbonyl nickel powder (×5000); (*a*) type 4; (*b*) type 123; (*c*) type 255.

merated particles having the form of chains or fibres. Figure 38 shows scanning electron micrographs of some forms of carbonyl powder. Similar variations in form can be obtained with powders produced by hydrometallurgical processes.

Wrought forms of pure nickel such as Nickel 200 show only the single-phase structure, the size and shape of the grains depending on the history of working and annealing. Electrodeposited nickel normally has a much finer grain size than the wrought and annealed metal, the structure depending on the conditions of deposition. With the high-speed nickel plating baths based on sulphamates (*see* Chapter 6, p. 121), the grain size increases with increasing current density.

Nickel–copper alloys

The nickel–copper alloys ranging from Monel alloy 400 (70 Ni, 30 Cu) to cupro-nickels (90 Cu, 10 Ni) are all basically solid solutions with grain structures depending on the working and thermal history. Monel K500 is hardened by the precipitation of $\gamma' Ni_3Al$, but the dispersion of the precipitate is too fine to be readily resolved. The alloy is more readily etched than alloy 400, but the structure is not significantly different (*see* Fig. 39), except that the grain boundaries are more clearly defined. The single-phase α nickel-silvers again show a similar appearance (*see* Fig. 40), with

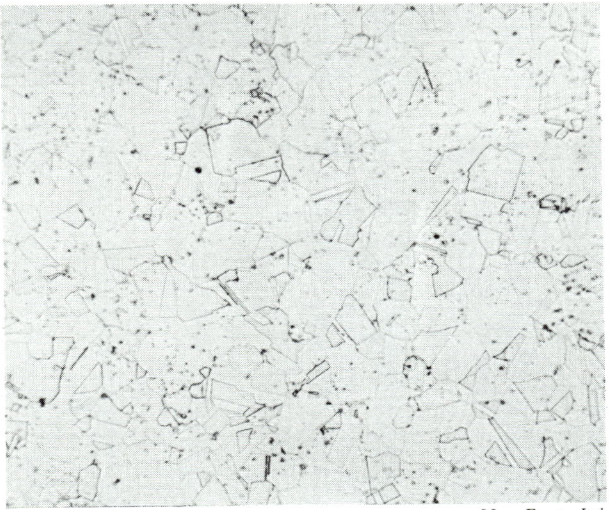

[*Inco Europe Ltd.*

Fig. 39—Monel alloy K 500, annealed ($\times$ 200).

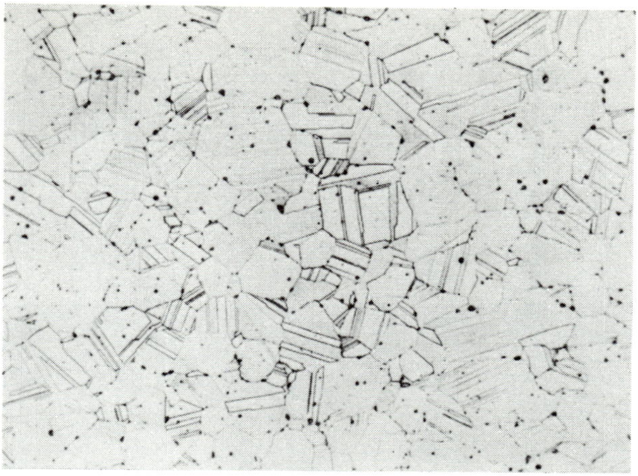

[*Inco Europe Ltd.*

Fig. 40—Nickel-silver α-phase alloy ($\times 200$).

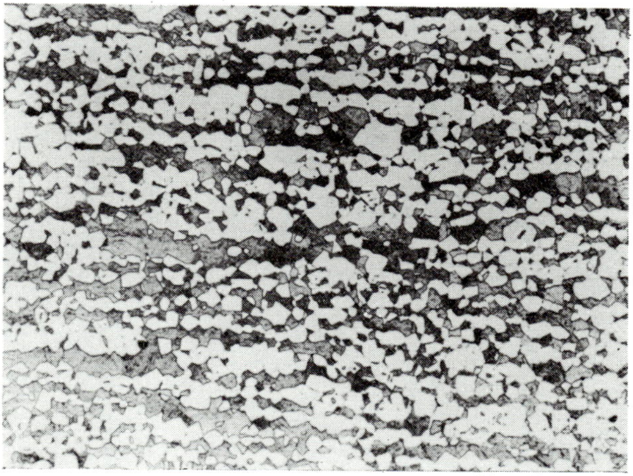

[*Inco Europe Ltd.*

Fig. 41—Nickel-silver α-β-phase alloy ($\times 200$).

perhaps a greater tendency to twinning than the binary nickel–copper alloys. The structure of the higher-zinc α-β alloys varies considerably with the proportions of the two phases. Figure 41 shows a typical structure.

NICKEL–CHROMIUM ALLOYS

The nickel–chromium binary alloys used for electrical heating-elements are also single-phase solid solutions, but traces of chromium carbides are normally visible, either at grain boundaries or within the grains, depending on the thermal history. However, the higher-chromium alloys developed to provide increased resistance to sulphur attack become two-phase with chromium contents above about 35 per cent.

The most interesting aspect of these alloys is the appearance of

(*a*)

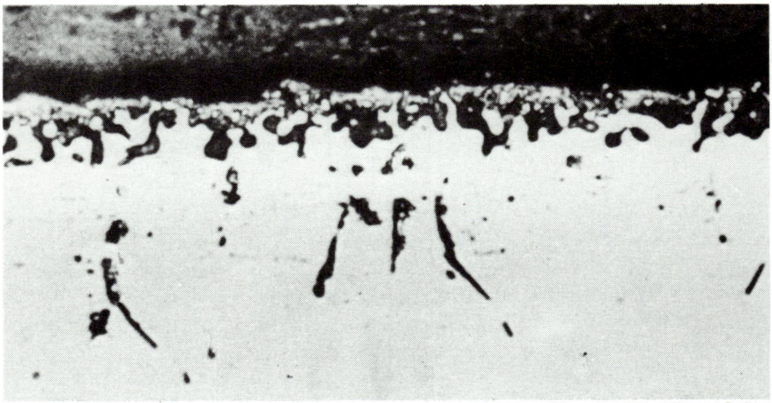

(*b*) [*Inco Europe Ltd.*

Fig. 42—Scale/metal interface on nickel–chromium alloys (×500): (*a*) without long-life additions; (*b*) with long-life additions.

the scale/metal interface after periods of high-temperature oxidation. Figure 42 shows the interface for two 80/20 nickel–chromium alloys, (*a*) with low contents of reactive elements and (*b*) with the higher levels of these elements adopted to make them suitable for intermittent heating. The interlocked interface in the latter case helps mechanically to key the scale to the metal.

Sulphur attack of nickel–chromium or nickel–iron–chromium alloys is another aspect of hot corrosion which develops characteristic metallographic features. Surface attack by sulphur at

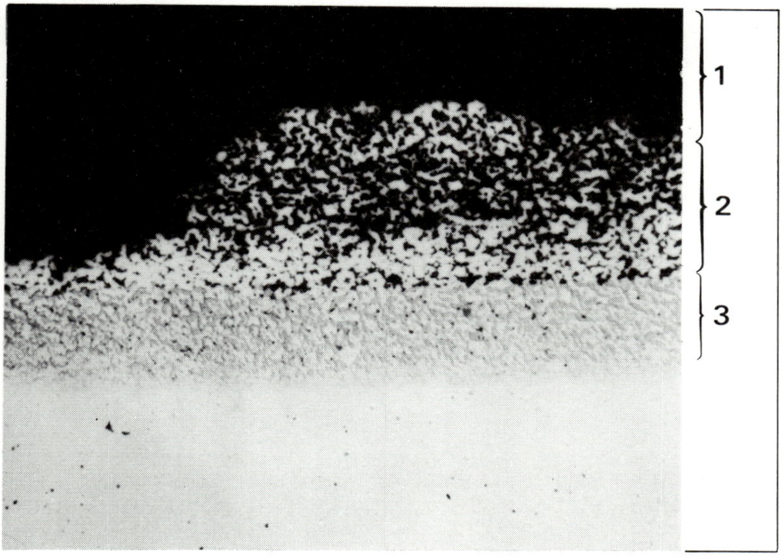

[*Inco Europe Ltd.*

Fig. 43—Hot corrosion of nickel–chromium alloy by sulphur-containing medium (× 100): 1. mount; 2. oxide/metal layer; 3. chromium sulphide layer.

temperatures in the region of 800–1000°C eventually leads to localised fusion due to the formation of the Ni–Ni_3S_2 eutectic, and produces wart-like excrescences on the surface (*see* Fig. 21). A microscopic examination of a section through such an attacked surface (*see* Fig. 43) shows at least two layers, the outer one consisting of chromium oxide dispersed within the residual metal phase, and the inner one showing the presence of the blue-grey chromium sulphide, Cr_3S_2. The metal is, of course, depleted in chromium, and near the surface is essentially only nickel. It is

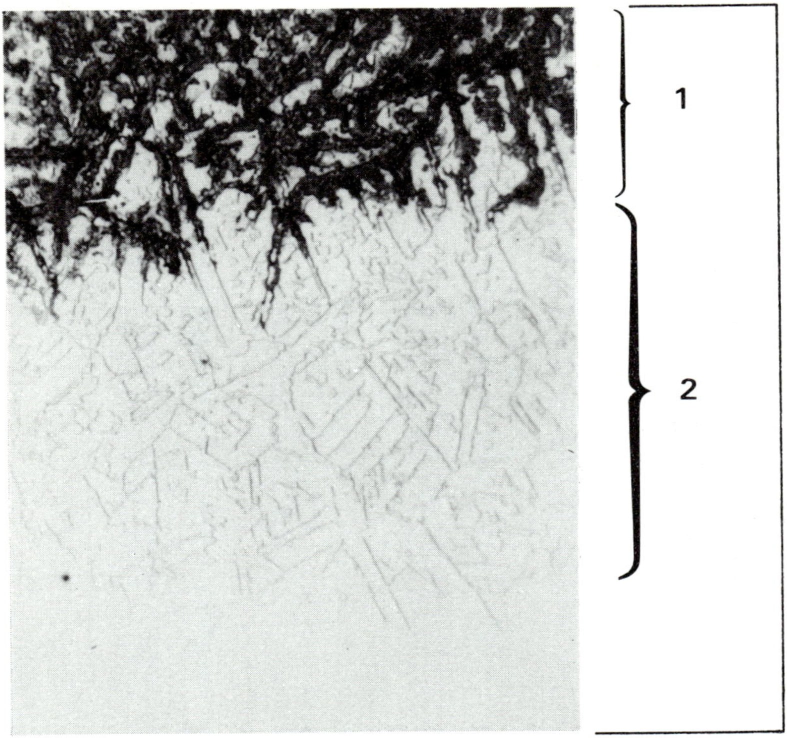

[*Inco Europe Ltd.*

Fig. 44—Hot corrosion of iron–nickel–chromium alloy by carbon-containing atmosphere giving "green-rot" attack (×1000): 1. oxidation of carbide; 2. chromium carbide.

hence susceptible to the formation of the low-melting-point eutectic on further attack by sulphur.

The inclusion of iron in the ternary nickel–iron–chromium alloys does not affect the appearance of the basic austenitic structure. Hot corrosive attack of alloys of this type by carbonaceous materials can lead to green-rot as described in Chapter 3. The formation of chromium carbides and the penetration of oxide along the needles of carbide is shown in Fig. 44.

Superalloys

Nickel-base superalloys show a wide variety of detailed microstructures depending on their particular composition and treatment, but they are particularly characterised by two features—the appearance of carbides and of the precipitated γ' hardening phase. Carbides are of two main types: the primary carbides, mainly titanium carbide, TiC (often combined with the nitride as carbonitride, Ti(C,N)), the appearance of which in Nimonic

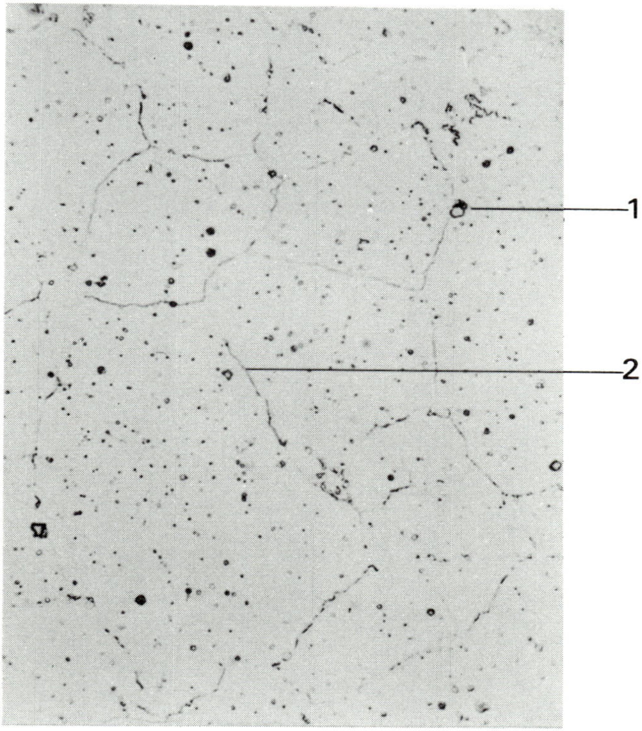

[*Inco Europe Ltd.*

Fig. 45—Nimonic alloy 75 showing primary and secondary carbides (×750): 1. Ti(CN); 2 chromium carbide.

75 is shown in Fig. 45; and secondary carbides precipitated either within the grains or, more importantly, at the grain boundaries. The primary carbides are visible in the unetched condition, varying from orange-yellow (nitride) to grey-mauve (carbide). Secondary carbides are soluble in the matrix at temperatures in the range 1100–1200°C and reprecipitate at lower temperatures.

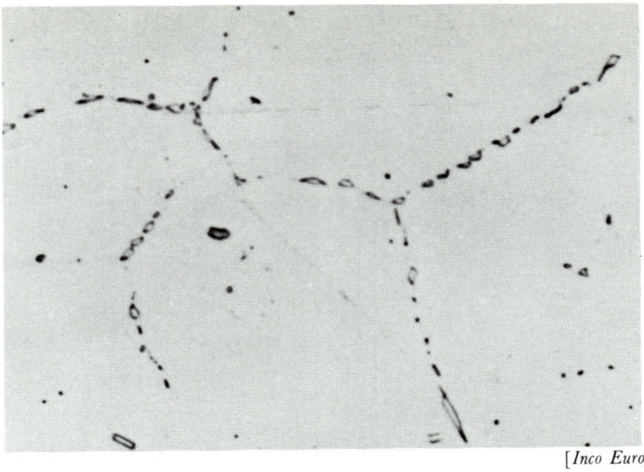

[*Inco Europe Ltd.*

Fig. 46—Grain-boundary chromium carbide in Nimonic alloy 80A ($\times 500$).

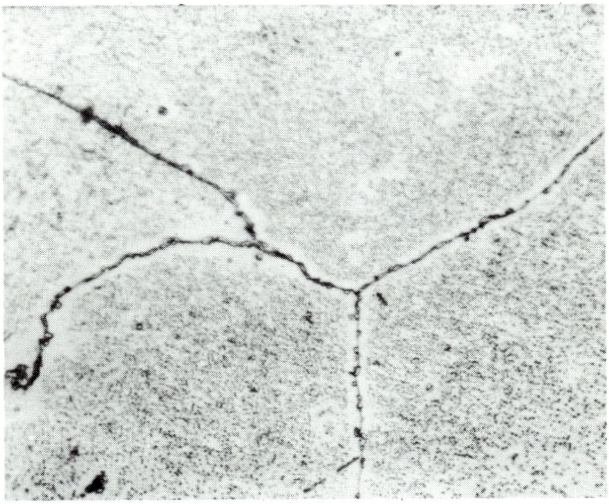

[*Inco Europe Ltd.*

Fig. 47—γ' precipitate in over-aged Nimonic alloy 90 ($\times 1000$).

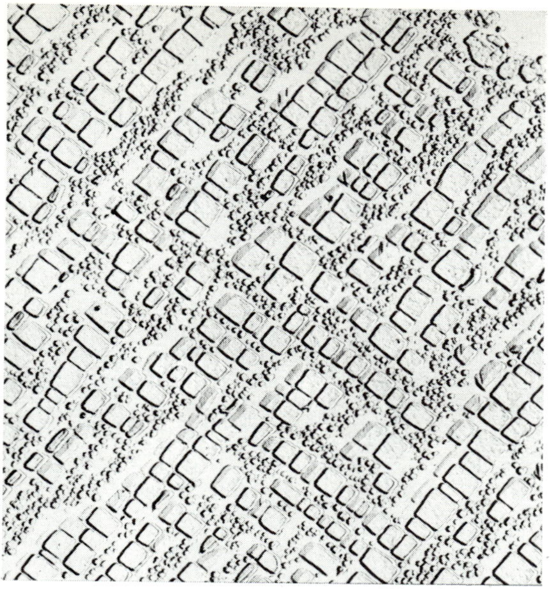

[*Henry Wiggin and Co. Ltd.*

Fig. 48—γ' precipitate in nickel-base superalloy ($\times$6000).

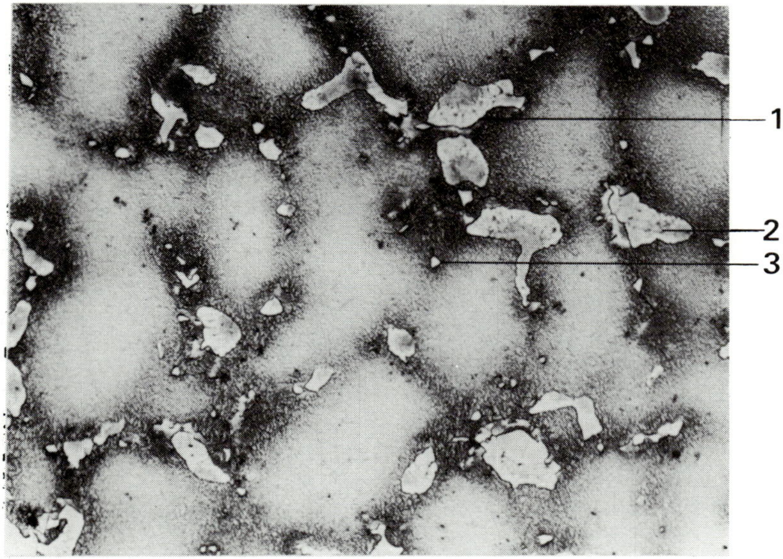

[*Inco Europe Ltd.*

Fig. 49—Cast nickel-base superalloy, M 20 ($\times$200): 1. γ' precipitate; 2. γ-γ' eutectic; 3. TiC.

They consist predominantly of chromium carbides, either Cr_7C_3 or $Cr_{23}C_6$. Figure 46 shows the appearance in Nimonic alloy 80A.

The γ' hardening precipitate is not resolvable optically in the normal condition of heat treatment, but can be coarsened by over-ageing. Figure 47 shows its appearance in Nimonic 90 after over-ageing for 24 hours at 900°C. The precipitate-free zones alongside the grain-boundary carbides are to be noted—they are important to the elongation of the material when tested under

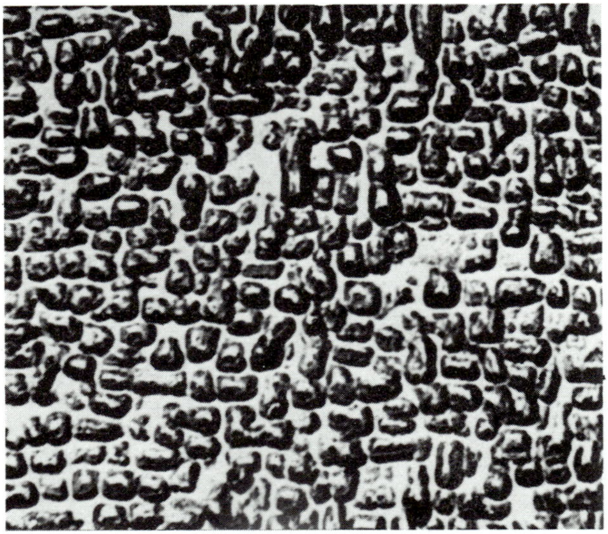

[*Permanent Magnet Association*

Fig. 50—Nickel–iron–aluminium permanent magnet alloy showing α-α' structure.

creep conditions. The structure of the γ' precipitate, and particularly the common orientation of the particles with respect to the matrix, is shown in the electron micrograph in Fig. 48. In the cast form, nickel-base superalloys show rather different structures, largely because of the higher levels of titanium and aluminium then commonly used. Figure 49 shows the structure of an alloy similar in composition to IN 100, in the cast condition.

Nickel–iron alloys

Most of the soft magnetic and controlled-expansion alloys of the nickel–iron system are single-phase alloys, but the permanent magnet alloys of the nickel–iron–aluminium type consist of duplex alloys with two isomorphous α phases—again the dispersion is normally too fine to be optically resolved, but the electron micrograph in Fig. 50 shows the oriented precipitate of the α' phase in the α matrix. The macro-structure of columnar permanent magnets in comparison with that of a similar alloy in the equiaxed form is shown in Fig. 33.

Reference

1. Smithells, C. J., *Metals Reference Book*, 4th edn., Butterworth, 1967.

CHAPTER FIVE

Production of Semi-fabricated Forms

The wide range of nickel-base alloys, and the number of forms in which such alloys are supplied, makes it difficult to give a concise but all-embracing survey of the methods adopted for their production. The melting point and strength of nickel lead, in many cases, to the use of procedures very similar to those in the steel industry. However, for other alloys, in which precise control of composition and freedom from contamination are essential, the techniques are more likely to be specially developed for the material in question.

MELTING

Although nickel and its alloys may be successfully melted in oil-fired, gas-fired or electric furnaces, the tendency nowadays is to restrict operations more and more to electric furnaces. This is because of their greater flexibility and the complete freedom from the danger of fuel contamination. For the latter reason, solid-fuel firing is avoided—sulphur being the main danger—but town gas, natural gas or low-sulphur oils (less than 0.5 per cent sulphur) are all satisfactory. For the general run of nickel alloys, electric-arc furnace melting is the most economical method and is carried out in units melting up to 10 tonnes of metal. At the other extreme in size, some alloys required commercially in relatively small quantities may be melted in crucible furnaces in amounts of only about 50 kg.

The general procedures are similar for most alloys[1]:

1. Use the purest raw materials consistent with satisfactory cost, avoiding contaminated scrap. All scrap must be cleaned and degreased before melting.
2. Avoid contamination during melting from refractories, slags, etc.

3. Reduce gas content by producing a carbon monoxide boil.
4. Deoxidise with silicon or aluminium.
5. Treat with titanium to fix nitrogen.
6. Control sulphur with magnesium or calcium additions.

For pure nickel the procedure might be to melt the charge of electrolytic or carbonyl primary nickel together with cleaned selected scrap in an open-hearth furnace lined with silica-alumina bricks. No slag need be added, but some would be formed by fusion of the refractories. An addition of low-sulphur charcoal is made to give a carbon content of about 0.25 per cent. When the temperature reaches about 1550°C, nickel oxide is added to produce the carbon monoxide boil, reducing the carbon content of the melt to 0.05–0.10 per cent and removing dissolved hydrogen. Additions of 0.3 per cent manganese, 0.03 per cent titanium and 0.05 per cent magnesium are then made prior to pouring the metal at about 1540°C; nickel-base hardener alloys are frequently used.

For nickel–copper alloys such as Monel 400, coreless high-frequency furnaces are ideal. A basic lining of magnesia or magnesia-alumina permits the use of basic slags, typically a mixture of equal parts of lime and fluorspar, for desulphurisation. The primary metals and clean scrap may be melted together with 0.1 per cent of nickel oxide, and brought to about 1550°C; the carbon monoxide boil may be produced by inserting the tip of a carbon electrode into the melt. When the boil ceases, the electrode is removed and the melt deoxidised with 0.1 per cent silicon. Final additions of 1 per cent manganese, 0.025 per cent titanium and 0.06 per cent magnesium are then made, and the melt poured at 1460–1500°C.

Cupro-nickel alloys may be melted in oil- or gas-fired furnaces, but electric furnaces—arc or induction—are preferred because of the higher-speed melting, minimising hydrogen pick-up. The charge should be melted quickly, raised to 1450–1500°C and a small addition of nickel oxide made to eliminate carbon and hydrogen. Initial deoxidation with manganese is followed by final deoxidation with magnesium, phosphorus or silicon. The metal is poured at about 1275°C for the 90/10 alloy and at about 1350°C for 70/30.

Nickel-silvers are melted in a generally similar manner to cupro-nickels. The copper is generally melted first, followed by the nickel; scrap is then charged with the zinc and the temperature allowed to rise until a zinc boil is obtained—this is effective in removing dissolved gases. Primary deoxidation is with manganese and final deoxidation with magnesium or phosphorus. Pouring temperature depends on composition, but for the α-alloys (20–28 per cent zinc) it would be as follows:

Ni content	°*C*
10	1180–1210
15	1200–1230
18	1220–1250

The simpler nickel–chromium and nickel–iron–chromium alloys are also preferably melted in induction furnaces in magnesia-alumina crucibles or linings. The primary metals and selected scrap are melted together, and at about 1600–1620°C additions of aluminium, titanium and magnesium are made before pouring at about 1580–1600°C. The alloys may also be melted in arc furnaces, thus permiting the use of oxygen lances for the reduction of carbon and other impurities picked up from scrap, although there is considerable loss of chromium as well.

The more complex nickel-base superalloys are melted either in induction furnaces in air, or in vacuum. The deleterious effect of a number of low-melting elements, for example lead, tellurium, bismuth, etc., on the high-temperature properties of these alloys makes it essential to avoid contamination through scrap or other sources, while vacuum treatment helps to remove such elements, as well as gases. Vacuum treatment may be introduced either as a secondary treatment after air melting, by transferring the molten metal to a vacuum degassing chamber; by complete vacuum melting and casting; or by consumable-electrode remelting, the electrodes for which may have been air-melted and cast or vacuum-refined before casting. The electrode remelting process may be by vacuum-arc remelting or by arc-melting under a refining slag—the so-called electroslag refining process. The general procedures followed for all such alloys are similar, the major constituents (nickel, chromium, cobalt, molybdenum, etc.) being first melted together, the hardening additions of titanium and aluminium then being made to the melt. Finally the malleablising additions of boron, zirconium, magnesium, etc., are made shortly before pouring. The pouring temperature for such alloys is in the range 1450–1550°C.

INGOT CASTING

The casting practice for high-nickel alloys is essentially similar to that employed with steel, although the greater solidification contraction requires larger feeder heads. Exothermic or insulating feeder heads, together with exothermic compound added to the exposed metal in the feeder head, are of particular value in ensuring a high yield of sound ingot. Cast-iron moulds are generally used, although water-cooled copper-faced moulds have been used for some purposes. The moulds may be dressed with mixtures such as lard oil and graphite, or rape oil and charcoal, which by flaming give a reducing atmosphere during casting. The moulds should be warmed before casting to ensure that they are perfectly dry.

Pouring into the ingot moulds may be by conventional top pouring through tundishes, or by bottom pouring into a cluster of moulds arranged around a central downgate. For nickel-base superalloys, Durville casting has been used. The casting machine involves a combined reservoir and mould; the metal is poured into the reservoir, allowed to remain static for a while, and then transferred to the mould by tilting the assembly slowly and smoothly, avoiding turbulence. Clean ingots with a smooth surface finish are thereby produced, suitable for hot working, particularly extrusion, with a minimum of preparation. Frequently all that is necessary is to saw off the feeder head.

For consumable-electrode remelting, the electrodes are produced by air- or vacuum-melting processes involving full control of composition and purity, any further refining being regarded as a bonus. In the vacuum-arc process the cast electrode is remelted by a d.c. arc passing between the base-plate of the mould and the electrode. Molten drops of the alloy fall into a pool on the base-plate where they progressively solidify. The quantity of molten metal at any time is small, and solidification occurs across a narrow interface. Thus segregation is limited, and since solidification is predominantly unidirectional, an ingot free from central unsoundness results.

The electroslag process is similar in principle, but the electrode is progressively melted by immersion in a molten-flux bath which is resistance-heated by a current passing between the base-plate and the electrode. Drops of molten alloy fall from the electrode through the molten flux, and by correct choice of flux certain refining operations may proceed. The refined droplets fall into a pool of molten alloy which is progressively solidified as in the vacuum-arc process.

Both processes yield sound ingots with minimum segregation and are particularly valuable in the production of large components in complex superalloys.

HOT WORKING

After sawing off the feeder heads and machining, if necessary to remove surface defects and adherent scale, the initial break-down of the ingots may be carried out by hammer forging, press forging or extrusion, or for the softer materials by direct hot rolling. In these initial operations the reduction, per blow or per roll pass, is kept to a low level to avoid intergranular cracking of the cast structure. Heating for hot working must be carried out with low-sulphur fuels, and natural gas, oil or butane are commonly used, the latter increasingly so. Fuel oils should contain less than 0.2 per cent sulphur, and gases less than 0.25 g/m^3. Temperatures range from 850 to 1250°C, depending on the alloy and the working operation.

For nickel and the more easily worked nickel–copper alloys, hot working commences at about 1150°C and may be continued to about 900°C or even lower. For nickel–chromium alloys, and particularly the superalloys, the increased high-temperature strength makes hot working a much more difficult operation, and for the most advanced alloys the permissible temperature range for hot working may be very narrow, for example between 50 and 100 degrees. Furthermore, because of the low thermal diffusivity of the alloys it is necessary to keep the time of contact between tool face and ingot to a minimum, to avoid severe surface chilling and the consequent risk of cracking. Quick-acting forging presses or extrusion presses are therefore the main equipment used for such work. A 1500-tonne forging press will handle a 2720-kg (6000-lb) ingot of a superalloy and reduce it to fine-grained billet suitable for further hot-working operations. Extrusion, which has been made feasible for the higher-strength alloys by the introduction of glass lubrication, enables cast and machined billets up to 300 mm in diameter to be extruded in a single operation to bar or shaped section, in sizes ranging from about 20 to 120 mm in diameter, and from 6 × 32 mm to 80 × 200 mm rectangular section. Tube shells are also extruded from hot-pierced or drilled billets for further drawing to seamless tubing. The extrusion temperature for most nickel-base materials lies in the range 1150–1200°C and the extrusion speed of the product through the die falls within the range 0.5–12 cm/s. Too low a speed would result in the billet falling in temperature before the extrusion was completed, while too high a speed would lead to incipient fusion and break-up of the product. For cupro-nickels, which are extruded to shells for drawing to seamless tubing, the extrusion temperatures recommended are: 90/10 cupro-nickel, 900–1000°C; and 70/30 cupro-nickel, 950–1050°C.

Nickel-silvers, being more easily hot worked than the nickel-rich alloys, are commonly direct hot rolled from the cast ingot. The optimum rolling temperature depends to some extent on the composition of the alloy but generally lies in the range 900–1000°C.

Further hot working of the alloys is carried out on rolling mills designed for rod, strip or plate production, and after the initial break-down referred to above, much heavier reductions per pass are possible provided the temperature of the work is maintained within the range appropriate to the alloy. The hot-worked product may then be used for forging or machining to the required form or may be used as the starting point for cold-worked materials.

COLD WORKING

The equipment and procedures used for cold working nickel and its alloys are very similar to those used for steel, the main difference being that, because of the virtual absence of a natural oxide film, lubrication becomes more important.

For the production of sheet, hot rolling would normally have reduced the material to a thickness of 2–2.5 mm. In this condition an oxide film covers the metal which, before cold working, is removed by pickling. For nickel, nickel–copper and nickel–iron alloys, a 10 per cent sulphuric acid bath with 2 per cent sodium nitrate is suitable. This is operated at 90–100°C, the scale being removed in about an hour. For chromium-containing alloys the scale is more difficult to remove and an initial treatment in a molten-salt bath helps. These baths are usually based on sodium hydroxide, and may be made reducing by the addition of sodium hydride, or oxidising by the addition of sodium nitrate. Treatment in the salt bath is usually for a few minutes at about 500°C and the material is then water quenched. It is finally pickled in 7 per cent nitric acid with 2 per cent ferric chloride at 60–70°C, or in 20 per cent nitric acid with 5 per cent hydrofluoric acid at 40–60°C.

The sheet is cold rolled on mills suited to the particular alloy. For pure nickel and the softer alloys, 2-high mills are satisfactory, and rolling is continued until the hardness has increased considerably. The sheets are then annealed at 800–900°C in a bright annealing atmosphere, usually of burnt town gas or burnt butane.

The harder nickel–chromium alloys are rolled in more powerful 4-high or Sendzimer mills, and since they cannot readily be bright annealed they are normally annealed in air and pickled to remove the scale. The rolling and annealing operations are repeated until the required gauge is obtained.

Wire drawing is carried out on a wide range of nickel-base materials, the details of procedure varying considerably from alloy to alloy. In general, the production of cold-drawn wire in all nickel alloys starts from hot-rolled rod about 7–8 mm in diameter. The softer materials are drawn with proprietary soap and oil lubrication, and generally present few difficulties. However, the nickel–chromium alloys, whether the simple binary alloys for electrical heating elements or the superalloys for high-temperature springs, present some characteristic problems. In a typical production procedure for these harder alloys, the hot-rolled rod is first annealed at about 1050°C in a protective atmosphere. It is pickled, first in sodium hydroxide and then in acid, and coated with lead by immersion in a molten bath. It is drawn through tungsten carbide dies to about 2.5 mm diameter, deleaded by pickling in 10 per cent nitric acid and strand annealed at 1050–1100°C in cracked ammonia. It is flash pickled, re-coated with lead and drawn further in a continuous 8-die drawing machine to about 1 mm diameter. After again de-leading and annealing, the wire is drawn further to the required gauge through diamond dies with mineral oil or water-soluble oil as lubricant in a continuous machine. It is finished by strand annealing in cracked ammonia.

Cold drawing of rod and tube from extruded or hot-rolled precursors is fundamentally similar in principle to the drawing of wire and involves the development of a balanced programme of cold working, annealing and pickling if necessary, suited to the alloy being handled. The procedures are readily applied to the softer nickel and nickel–copper alloys but are also applied to nickel–chromium and nickel–iron–chromium alloys, and to the less highly-alloyed nickel-base superalloys.

FINISHING AND INSPECTION

The finishing of hot- or cold-worked products may involve heat treatment, machining and inspection (the procedures adopted with nickel-rich materials were described by Betteridge and Cound[2]). The heat treatments may be either annealing of cold-worked materials to make them suitable for further working operations, or hardening by solution and precipitation treatments to improve the mechanical properties.

Annealing treatments for nickel and its alloys in the form of sheet or strip are at about 800–1050°C, an average treatment time being about 15 minutes. The treatments may be in batch furnaces in air, but preferably would be carried out in continuous furnaces with a protective atmosphere. In the most widely-used furnace of this type the work is carried on a continuous woven-wire belt through the heating zone and a water-jacketed cooling zone, with an atmosphere of burnt town gas, burnt butane or cracked ammonia.

The final annealing of wire products is normally carried out by strand annealing. The wire is passed through electrically-heated heat-resisting metal tubes at speeds up to several metres per second in a suitable reducing or protective atmosphere. The annealing time in such cases may be less than one second, the temperature being therefore higher than that used in batch or conveyor-type furnaces. In particular, for nickel–chromium wires the temperature is 1050°C. Strip is annealed similarly, although the speeds of travel are usually very much lower.

Precipitation-hardening treatments of all metals need to be carefully controlled in terms of temperature, time and cooling rates in order to achieve the required properties. Details of the structures developed were given in Chapter 4 for many of the alloys. The only item of general applicability that should be mentioned here is to emphasise again the importance with nickel alloys of avoiding contamination by sulphur-containing materials. Fuels and refractories should have low-sulphur contents and care should be taken to remove all residual cutting oils or lubricants from the material before heat treatment.

Inspection of nickel alloy products for internal defects is mainly restricted to two techniques—ultrasonic and eddy-current. The former is mainly of value in examining relatively large articles such as superalloy rotor discs and billets at intermediate stages of working, but is also applied to small machined blanks of superalloys prior to forging into turbine blades. Eddy-current testing has proved of particular value in the examination of bar and tubing in nickel-base alloys. The method is sensitive to both surface and internal defects and is therefore best applied to products with a machined surface on which surface defects have been reduced to a minimum.

Visual inspection for surface defects is best carried out after pickling or electrolytic etching. The macro-etching reagents given in Chapter 4 are generally suitable for inspection purposes, and apart from revealing the grain size of most materials, will show up surface defects arising from working operations.

In many cases alloys must meet particular requirements with regard to physical and mechanical properties, or be capable of being processed by particular manufacturing operations. A variety of routine tests for such characteristics as magnetic properties, electrical properties, expansion behaviour and normal-temperature or high-temperature mechanical properties are carried out on nickel-containing alloys. In addition, special tests have been devised for many materials to confirm their suitability for particular processes; these embrace tests for weldability, for wire-spiralling behaviour, for gas content and for plating characteristics.

CASTINGS

Although by far the majority of all nickel-base alloys is used in the wrought form, there has always been a significant use of many of the alloys as castings. Thus nickel, nickel–copper alloys and nickel–chromium alloys, usually with minor modifications of composition in comparison with the corresponding wrought alloys, have been used as sand castings for a variety of components of complex shape which would be difficult to machine or fabricate. In more recent years, with the introduction of special alloys, particularly in the superalloy class, castings made by one of the precision-casting techniques have become more widely used.

Foundry techniques in general are more akin to those used for steel than for other non-ferrous metals. Dry sand moulds with high permeability are preferred for pressure castings or for thicker section castings, although green sand moulds may be satisfactory for thinner sections. Plaster moulds have been used for nickel-silver to give a smoother surface finish requiring less finishing. The shrinkage allowance for nickel alloys is 2 per cent. The alloys usually have higher silicon contents than the corresponding wrought alloys. Table 29 gives typical compositions. The higher silicon versions of cast Monel have increased hardness and wear resistance (*see* Chapter 3, pp. 38–41) and at about 4 per cent silicon are precipitation hardenable. The alloys are melted and deoxidised by the methods already described and generally are poured at temperatures within the range 1450–1560°C, depending largely on the section and complexity of the casting.

Cupro-nickel alloys and nickel-silvers, because of their lower melting points, are cast at lower temperatures, usually in the range 1250–1400°C, again depending on the composition and on the complexity of the casting.

Nickel–chromium alloys are not good foundry alloys because of the tenacious oxide film formed on the molten alloy and the relatively poor fluidity. Nevertheless, with careful mould design, using generous section runners and ensuring smooth entry of the metal into the mould, satisfactory results can be obtained. Dried silica-sand moulds, carbon dioxide sand and resin-bonded shell moulds have all been used successfully with nickel–chromium alloys. The use of insulating or exothermic sleeves or feeder heads helps to

TABLE 29

Typical compositions of cast nickel alloys.

Alloy	*Ni*	*Cu*	*Cr*	*Fe*	*Mn*	*Si*	*C*	*Specification*
Nickel	97	0.3	—	0.25	0.5	1.5	0.5	—
Monel	63–60	32	—	1.5	0.75	1.6–4.0	0.15	BS 3071: NA 1 1% Si NA 2 2.75% Si NA 3 4% Si
Inconel	75	0.25	16	6.0	0.8	2.0	0.2	—
50 Ni : 50 Cr	48	—	50	1.5	0.5	0.3	0.1	—

For cast nickel-base superalloys, *see* Table 17.

promote good feeding of the castings. Techniques of this type enable satisfactory castings for general heat-resisting purposes to be made and have been applied to nickel–chromium and nickel–iron–chromium alloys containing 15–25 per cent chromium, as well as to the high-chromium binary alloys containing 50–60 per cent chromium.

The nickel-base superalloys used in the cast form are mainly cast by precision methods which ensure that very little subsequent machining is required. In the lost-wax technique, a wax pattern is given a fine refractory coating and is then invested in a refractory slurry bonded, for example, with ethyl silicate. After drying, the wax is removed by melting, and the mould heated to about 1000°C for casting. The simpler superalloys may be melted and cast in air, but the most highly-developed alloys with the best high-temperature strength are melted and cast in vacuum. By this means precision castings are obtained in sizes ranging from small turbine-blades (say 2 cm long) to complete rotors up to 40 cm or more in diameter.

POWDER METALLURGICAL PRODUCTS

Pure nickel powders may be produced by various methods, but the principal ones in commercial use are by the controlled decomposition of nickel carbonyl or by ammonia-leach hydrometallurgy, both of which have been referred to in more detail in Chapter 2, pp. 24–25. The powders may be used to produce pure nickel articles, or combined with other metals to produce alloy parts, and these may be either fully dense or porous[3]. The advantages of using powder techniques to produce fully dense articles are that high purity can be maintained more readily than in melting, and that more precise alloy compositions can be achieved; this is of particular importance for magnetic and controlled-expansion alloys.

To consolidate pure nickel powder it may be pressed unidirectionally or isostatically, or it may be compacted by continuous rolling. For unidirectional pressing the powder is fed into steel moulds and a high pressure applied through a punch. Figure 51 shows the relationship between pressure and density of the green compact for the standard grade of carbonyl nickel powder. The compact is sintered, typically for one hour in dry hydrogen at 1000°C, where its density increases. For a compact pressed at 420 MN/m^2, the sintered density reaches about 8300 kg/m^3, corresponding to a porosity of about 6.5 per cent. Fully-dense products can then be obtained from such compacts by conventional hot and cold rolling. Isostatic compaction is carried out by enclosing the powder in a rubber bag from which the air is extracted, and compressing the assembly in a fluid medium; the compact is then sintered and further processed as a unidirectional pressing. Continuous-roll compaction to strip is carried out by feeding the powder at a carefully-controlled rate into the gap between rotating rolls; spiky powders of the type shown in Fig. 38(*c*) are preferred for this operation in order to obtain sufficient strength in the green strip, which typically might have a porosity of 28 per cent. The green strip is sintered in a reducing atmosphere for a short time at 1100–1200°C, reducing the porosity to about 10 per cent, and is then successively cold rolled and annealed to achieve complete densification.

Porous nickel articles such as battery plates or fuel-cell electrodes can be produced by similar processes using either pressureless or low-pressure compaction to give the required degree of porosity. In such cases initial sintering is carried out with the powder retained in the mould and a reinforcing mesh or grid is usually incorporated. For very high levels of porosity a volatile spacing agent such as methyl cellulose can be added which is subsequently removed by heating to 200–300°C. Continuous-roll compaction may also be used for the production of porous strip, and slurry methods have also been used.

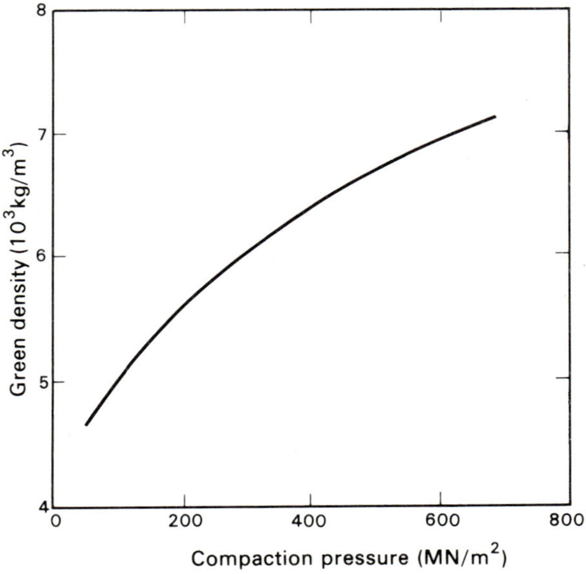

Fig. 51—Relationship between pressure and density of unsintered nickel powder compacts.

Powder metallurgical methods are being used for the production of shaped articles in nickel-containing alloys, particularly nickel-iron alloys. Isostatic compaction is mainly used for the consolidation of billets for further processing, but unidirectional pressing is the chief process used for producing finished components.

Powder techniques may also involve the use of pre-alloyed powders in addition to, or instead of, elemental powders. For this process the alloys are melted in the usual manner and the liquid alloy is blown to powder by a suitable gas jet. The powder is then compacted and sintered in a similar manner to that described for elemental powders. The advantages sought for in this process are the avoidance of segregation and other defects inherent in the solidification of large ingots from the molten state, and overcoming the difficulties of homogenisation which would arise with complex alloys if they were compounded from elemental powders. The technique is showing particular promise in the production of large rotor discs in nickel-base superalloys.

The dispersion strengthening of nickel and nickel alloys by the inclusion of a small percentage of inert particles, usually refractory oxides, necessarily involves powder metallurgical processes.

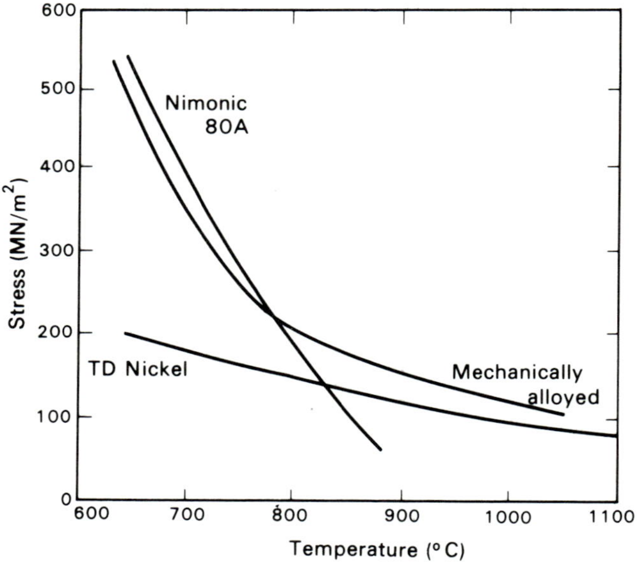

Fig. 52—100-hour stress-rupture strength of mechanically-alloyed nickel-base superalloy in comparison with Nimonic 80A and dispersion-hardened TD nickel.

The usual dispersant is thoria and it may be introduced by direct addition of thoria to the loose metal powder and milling; by co-precipitation of nickel and thorium oxides and selective reduction of the former to metal; or by wetting nickel powder with a solution of a thorium salt, drying and thermally decomposing the salt to oxide. Subsequent processing is analogous to that for the pure metal. Dispersion hardening has recently been extended to nickel-base superalloys by the process termed mechanical alloying[4], in which mixed elemental nickel and chromium powders, a master-alloy powder containing the titanium and aluminium, and a dispersant of yttria, are milled together by a dry technique in a high-energy ball mill. This milling process promotes particle welding and attrition, the product being then consolidated by extrusion in evacuated, sealed containers. The final extruded bar combines the merits of γ'-hardened and dispersion-hardened high-temperature alloys, as shown in Fig. 52, in which the 100-hour rupture stress for Nimonic 80A, for TD Nickel, and for a mechanically-alloyed material are plotted against temperature.

References

1. Bieber, C. G., and Decker, R. F., "Melting of Malleable Nickel and Nickel Alloys", *Transactions of the Metals Society of the American Institute of Mining and Metallurgical Engineers*, vol. 221, p. 629, 1961.
2. Betteridge, W., and Cound, T. E., "The Heat-treatment, Inspection and Testing of Wrought Nickel and Nickel Alloys", *Journal of the Institute of Metals*, vol. 83, p. 262, 1955.
3. Tracey, V. A., "The Properties and Some Applications of Carbonyl Nickel Powders", *Powder Metallurgy*, vol. 17, p. 54, 1966.
4. Benjamin, J. S., "Dispersion Strengthened Superalloys by Mechanical Alloying", *Metallurgical Transactions*, vol. 1, p. 2943, 1970.

CHAPTER SIX

Processes Applied in Final Manufacturing

The processes described in the previous chapter yield what are termed semi-finished products, which are those normally supplied by metal manufacturers to users. In this chapter processes are described which are normally applied by users to shape the metal to the required form and to finish the surface.

FORGING

Hot forging, either between flat hammer faces or with open dies, follows very much the lines described previously for ingot forging. The temperatures used for the different alloys are similar to those previously mentioned, and precautions to avoid contamination by sulphur must be maintained. It is important to guard against the use of insufficient power to cope with the work in hand, since this can lead to low levels of deformation, frequent reheating, and the growth of coarse grains. Particularly with nickel-base superalloys, for which the creep and thermal fatigue properties are dependent on controlled and uniform grain size, this is an important factor. Forgings in these alloys range in size from small gas-turbine blades weighing only a few hundred grams to rotor discs up to 450 kg in weight.

Blade forgings fall into two categories: oversize forgings with 1–2 mm excess on the aerofoil section; and precision forgings 0.1–0.25 mm oversize. The former type are fully machined after forging, while the latter are only polished. The forging operation falls into two stages, the first being the production of a "use" in which the length of bar is thickened at one or both ends to provide the bulk of metal for the root and the shroud of the blade. This operation may be carried out by electrical or mechanical upsetting of the ends, by partial extrusion, or by "necking" a larger-diameter bar. The second stage is the die-forging of this use to blade form. Usually at least two heatings are required, the first to produce the rough blade-form and enable excess metal to be removed by cutting off the flash, and the second to produce the final form in a single blow in the finishing die. Lubrication of the dies is necessary, commonly by a graphite suspension, although glass or nickel plating may also be used. Nickel plating is of particular value in minimising intergranular oxidation of the alloy in the production of precision forgings.

For the larger disc forgings, extremely powerful hammers or presses are required. The starting material is a hot-worked billet. This is first forged to an appropriate cheese or pre-form shape before die-forging, the object being to ensure adequate metal flow in all parts of the final forging. Several reheating operations may be required.

COLD FORMING

Nickel and its alloys more closely resemble steel in their cold-working behaviour than other non-ferrous metals. However, careful attention must be paid to the choice of lubricants, because of their thinner oxide film than that which helps prevent metal-to-metal contact in mild steel. The light oils and water-soluble lubricants suitable for drawing steel are not usually sufficient, and heavier-duty oils containing surface-active constituents or inert filler materials are preferable.

Tool and die materials need to be hard and wear resistant if long life is required. Plain carbon steels are not recommended because of their tendency to gall with high-nickel alloys, but most other standard materials are satisfactory. Hard alloy bronzes, nickel–chromium cast iron, chromium-plated hardened steel and tungsten carbide have all been used successfully.

The softer, more ductile, materials (nickel, nickel–copper alloys, nickel-silver, etc.) are amenable to most of the cold shaping operations such as deep drawing, spinning, etc., and these operations can also be applied to the simpler nickel-chromium and nickel–iron–chromium alloys.

MACHINING

The behaviour of nickel and its alloys in machining again resembles that of steel rather than other non-ferrous metals.

However, their toughness demands that attention must be paid to a number of points[1]:

1. Maximum rigidity of both tool and work.
2. Sharp tools with high surface finishes.
3. Clearance angles of tools no larger than necessary.
4. Copious supply of cutting fluid.
5. High feed-rate to minimise work hardening.
6. Maximum clearance of swarf.

The more ductile materials such as nickel, nickel–copper alloys and nickel-silver are frequently more satisfactorily machined in the cold-worked rather than the annealed condition, particularly when the stock is small-diameter rod which is liable to bend under the side pressure of the tool. Age-hardenable materials may be machined either in the solution-treated condition or after full heat treatment, the choice often depending on the particular operation involved.

The tool materials may be high-speed tool steels or tungsten carbide tips. For most general purposes with the softer materials, an 18 per cent tungsten, 4 per cent chromium, 1 per cent vanadium steel is satisfactory, but for heavier duty a cobalt-containing steel of the same type is used. For materials with hardnesses over 280 HV carbide-tipped tools are usually preferred, but rigidity of the whole set-up is essential, and their use should be confined, if possible, to continuous rather than intermittent cutting operations.

The tough nickel alloys develop high frictional heat during machining and hence cutting fluids are necessary for the best results. Water-soluble cutting oils are generally satisfactory, but for the stronger, tough alloys, such as the nickel–chromium superalloys, a straight extreme-pressure cutting oil may be preferred. All lubricants must be thoroughly removed before subsequent heat treatment.

Apart from the normal mechanical methods of machining, a number of new chemical or electrical methods have been introduced which are of particular value in dealing with high-strength superalloys. Electrochemical machining is essentially the reverse of electroplating—the workpiece forms the anode and a shaped cathode, usually of copper, is used at a distance of a few millimetres or so in a suitable electrolyte. This electrolyte is usually a neutral salt solution or dilute acid, from which the dissolved metal is readily precipitated as insoluble salts to prevent deposition onto the cathode. The process is independent of the mechanical properties of the alloy and hence can be applied to fully heat-treated components. A development of this process is electrolytic grinding, in which a conductive, abrasive wheel is used in contact with the workpiece. Any passive film is ground away, but electrochemical action is responsible for the major proportion of the metal removal.

Spark erosion is based on the passage of an electrical discharge between a copper counter-electrode or tool, and the workpiece, while immersed in a dielectric fluid. The counter-electrode is oscillated to make intermittent contact with the workpiece and metal is removed by the thermal energy of the discharge. The removed metal forms a fine sludge in the dielectric. This process is mainly applied to the drilling of deep holes or recesses in hard superalloys.

WELDING, BRAZING AND SOLDERING

Joining processes of many types are available and, although they are not all applicable to the complete range of nickel alloys in all sections, there is usually no difficulty in selecting one suitable for any particular requirement[2]. The more ductile solid-solution alloys are the more readily handled, and to these almost all welding processes have been successfully applied at some time. These materials should preferably be welded in the annealed condition, although moderate cold-forming operations such as tube bending do not necessarily involve re-annealing. Precipitation-hardening alloys introduce difficulties because of the localised response of the alloy to the heat generated in welding; these alloys are normally welded in the solution-treated or solution-annealed condition, and either fully heat-treated or precipitation treated after welding. The thickness of weldable section in these alloys may be restricted and the majority of welding on them is carried out by the tungsten-arc inert-gas process or by resistance-welding techniques.

For all alloys it is important that the surfaces surrounding the joint are cleaned immediately before welding is attempted. Mechanical abrasion with wire brush or grinding wheel should be extended to at least 2 cm from the joint edges and should be followed by chemical or solvent degreasing. Adequate jigging, not only to maintain the parts in their relationship, but also to restrict distortion, is necessary, especially for thin sheet components. Tack welding prior to the main runs is helpful.

Nickel and nickel–copper alloys may be welded by the oxy-acetylene torch, but most work is done by the metal-arc process using coated electrodes, which are available in suitable matching compositions. Argon-shielded tungsten-arc and metal-arc processes are also applicable and filler wire of matching composition is used. Overlaying of steel for corrosion protection is common practice, using nickel or nickel–copper alloys, and the metal-arc or argon-shielded metal-arc processes are followed. When coating steel with nickel–copper alloys, the amount of iron taken up by the alloy coating must be restricted to prevent embrittlement. This may be effected by control of the welding conditions, or by making an initial overlay of nickel before depositing the alloy. The welding products, electrodes and filler wire for these alloys include additions of elements to control the gas content of the weld metal and thus to reduce porosity in the welds; these additions are usually silicon, titanium and sometimes aluminium.

The solid-solution alloys of the nickel–chromium and nickel–iron–chromium systems are also readily welded by most of the conventional processes. Appropriate consumable materials of matching composition are available, although there is a tendency to use consumables based on 80 nickel–20 chromium even with the iron-containing parent materials. With chromium-containing materials the fluxes used are boron-free, since boron is liable to cause hot cracking in the weld deposit.

Precipitation-hardening alloys, particularly the γ'-hardened superalloys, are usually welded by the inert-gas tungsten-arc process, using matching filler wire, although below about 2 mm in thickness butt welds can be made without filler, providing the edges match closely. An a.c. or d.c. arc may be used, but generally d.c. is preferred with the tungsten electrode negative. Argon and argon–hydrogen are the most commonly used shielding gases and the supply needs to be generous to avoid oxide inclusions in the welds. Figure 53 shows a section of a multi-run inert-gas arc weld using matching filler wire in a nickel-base creep-resisting alloy.

Under favourable conditions the strength of welds made autogenously or with added matching filler material may be close to that of the parent material, as shown in Table 30, which gives the tensile properties of welded joints in Nimonic 90 sheet (after appropriate post-weld heat treatment) in comparison with those of the unwelded sheet. In some cases where the closest match in mechanical properties is not necessary, it may be advantageous to use a more ductile filler material to reduce the risk of weld-metal cracking. The weldability of the γ'-hardened nickel-base high-temperature alloys is dependent on the level of titanium and aluminium in the alloy. Thus Nimonic 80A and Nimonic 90 may

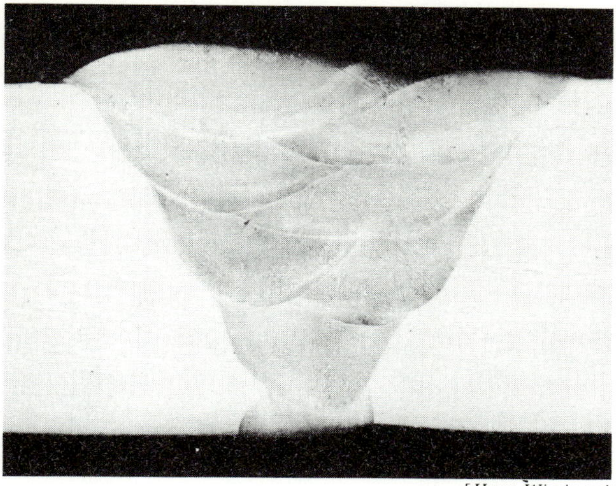

[*Henry Wiggin and Co. Ltd.*

Fig. 53—Multi-run inert-gas arc weld in nickel-base superalloy using matching filler wire ($\times 3\frac{1}{2}$).

be tungsten-arc welded satisfactorily in thicknesses up to about 5 mm, but Nimonic 115, for example, is regarded as unweldable by this process. Special welding processes such as plasma-arc and electron-beam methods extend the scope for these alloys, but the low high-temperature tensile ductility of the alloys makes them very susceptible to cracking in the heat-affected zone.

Electrical-resistance welding may be used with almost all nickel-base alloys if due attention is paid to their special characteristics. These necessitate particularly careful cleaning of the chromium-

TABLE 30

Tensile properties of welded Nimonic 90 sheet.

Temperature	*Parent material*			*Welded and heat-treated sheet*		
(°*C*)	*0.2% proof stress* (MN/m^2)	*Tensile strength* (MN/m^2)	*Elongation* (%)	*0.2% proof stress* (MN/m^2)	*Tensile strength* (MN/m^2)	*Elongation* (%)
20	794	1115	17	831	1180	23
600	709	925	13	729	961	23
700	714	820	4	703	848	8
800	563	604	2	548	669	6

containing alloys to avoid poor electrical contact between the mating surfaces, and the use of higher electrode pressures than with steel because of the strength retained at elevated temperature. Flash welding has been used with nickel-base superalloys to produce ring structures from extruded section. The process involves striking an arc between the prepared faces of the mating pieces

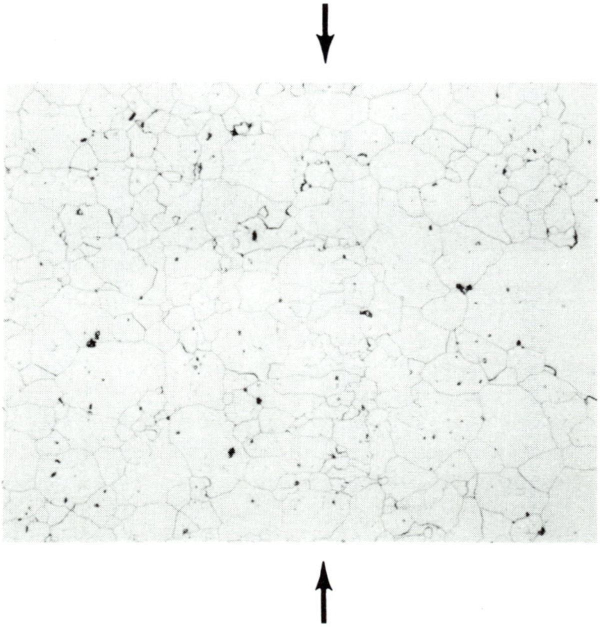

[*Henry Wiggin and Co. Ltd.*

Fig. 54—Section of flash-butt weld in Nimonic alloy 90. Note the virtual absence of weld-affected metal. The line of weld is between the two arrows. ($\times 100$.)

and its maintenance until the faces are molten. The pieces are then forced together to expel all liquid, giving a joint with strength properties essentially those of the parent metal. Full heat treatment is desirable after welding. Figure 54 shows a section through such a weld in Nimonic alloy 90.

Soldering or brazing may be used to join nickel alloys to themselves or to other materials, or to seal mechanically-locked joints. A very wide range of joining alloys may be serviceable, ranging from ordinary tin–lead soft solders, applicable mainly to nickel,

nickel–copper alloys or nickel-silvers, to high-temperature brazing alloys, some of which are suitable for joining nickel-base super-alloys for service at temperatures as high as 800°C. The strengths of such joints are, of course, low in comparison with that of the parent metal, but by suitable design, using relatively large areas of mated surface (in comparison with the ruling sections of the individual parts), adequate strength can be obtained.

The effectiveness of all soldering and brazing operations depends very largely on the type of flux used. For soft soldering of nickel and nickel–copper alloys, an active flux based on "killed spirits" (excess of zinc added to hydrochloric acid) or zinc chloride is recommended. A suitable composition is as follows:

zinc chloride	100 g
ammonium chloride	100 g
glycerine	100 g
water	1 litre

A small amount of a wetting agent may usefully be added.

For chromium-containing alloys, a still stronger flux is required, and killed spirits with the addition of 10 per cent acetic acid are suitable. Such fluxes are very corrosive and all traces must be removed from the article after soldering, by careful washing.

Silver soldering or high-temperature brazing uses alloys with melting points ranging from 630°C for alloys of the silver–copper–zinc–cadmium system to over 1200°C for some nickel or palladium-base alloys. Table 31 lists a selection of these alloys which are useful in joining nickel-base alloys either to themselves or to other alloys. In applying these to the joining of nickel-base alloys it is important, especially when using the lower melting-point silver solders, to ensure that the components are stress free, i.e. they must be annealed and not stressed in assembly; otherwise there is serious danger of intergranular penetration by the solder. Nickel, nickel–copper and nickel–iron alloys may be brazed in an inert or reducing atmosphere or in vacuum. However, with chromium-containing alloys, and especially those with additions of titanium and aluminium, an active flux is necessary. There are a number of effective proprietary brazing fluxes available, mostly based on borax. The quantity of flux used should be kept to a minimum because of the possibility of intergranular attack by a boron eutectic, and in some cases it is found advantageous to first pickle the parts and then plate with copper or nickel before brazing. Clearances to be filled with the brazing alloy normally lie in the range 0.1–0.5 mm (radial or gap). Closely-fitting parts, such as

TABLE 31

Silver solders and high-temperature brazing alloys used with nickel alloys.

Alloy	*Melting range* (°*C*)	*Brazing temperature* (°*C*)	*Specification*
Ag–Cu–Zn–Cd	620–640	650	BS 1845/3
Ag–Cu–Zn	690–735	740	BS 1845/4
Ag–Cu–Pd	807–810	815	BS 1845/Pd 1
Ag–Cu–Pd	900–950	955	BS 1845/Pd 6
Ag–Pd–Mn	1000–1115	1120	BS 1845/Pd 9
Pd–Ni	1237	1250	BS 1845/Pd 14
Ni–Cr–B–Si	955–1000	1020	BS 1845/Ni 3
Ni–Si–B	960–1040	1050	BS 1845/Ni 4
Ni–Mn	995–1080	1100	—
Au–Ni	950	950	BS 1845/Au 6

are commonly copper-furnace brazed when in steel, are unsuitable since the brazing alloys are to some extent soluble in the nickel alloy. It is appropriate to mention here that while nickel alloys can be furnace brazed with copper, the solution of nickel by the copper raises its melting point and hinders the capillary flow of the brazing medium, the above comments on clearances being therefore relevant. For high-temperature brazing and soldering, heat may be applied by any of the conventional methods—resistance-furnace, induction, electric resistance, torch or salt bath—and the brazing alloy may be pre-placed as wire or powder, or fed to the joint during the brazing operation.

The strengths of joint obtained by brazing processes depend to a great extent on the design of the joint and on the efficiency with which the operation is carried out. Stress-rupture properties, usually measured with the joint stressed in shear, have shown that the strongest brazing alloys may give joints with about 10 per cent of the strength of a nickel-base superalloy at temperatures in the range 600–800°C; at this level, joints giving useful service life may be designed.

ELECTRODEPOSITION

One of the major uses of nickel is an electrodeposited coating for purely decorative purposes, as a corrosion-resisting coating or as a hard wear-resisting surface. Electrodeposited nickel is also being increasingly used as a self-supporting article after stripping from the former onto which it was deposited. This process is called electroforming. The principles of electrodeposition are well established, and it is sufficient to indicate here that if an electrical potential is applied to two electrodes immersed in a solution of a nickel salt, the nickel ions, being positively charged, will be attracted to the cathode, where they give up their charge and are deposited as nickel atoms. At the same time, the negative ions travel to the anode, and if this consists of metallic nickel, the discharged ion will combine with the nickel of the anode and take it into solution in the electrolyte. Thus the concentration of salt in the solution is maintained, and metallic nickel is transferred from the anode to the cathode.

Faraday's law indicates that the passage of one ampere for one hour would deposit 1.095 g of nickel if the process operated at full efficiency. Most nickel-plating electrolytes actually operated at about 97 per cent efficiency, the remaining current evolving hydrogen at the cathode. This means that in practice about 1.05 g of nickel are deposited per ampere-hour.

The control of electrodeposition to ensure that consistent results (meeting the required characteristics) are obtained involves: choosing anode material and locating it in the bath according to the work being plated; choosing the electrolyte and maintaining it; and choosing the operating conditions such as temperature, current density, electrolyte stirring, etc. Only the essentials can be outlined here and the reader is referred to standard textbooks for more detailed guidance[3].

Nickel plating

In order to ensure smooth and uniform dissolution of the anodes in the electrolyte, activated forms of nickel are commonly used. These consist of nickel with minor additions of other elements, usually sulphur or oxygen, which prevent anode passivity; they are termed depolarised anodes. Carbon-containing anodes are used in bright nickel-plating baths to minimise the formation of fine metallic sludge which would lead to roughness in the deposit as well as reduced anode efficiency. The carbon, which is uniformly distributed through the anode, forms a continuous and tenacious sheath on the surface of the anode during dissolution and this serves to retain the metallic sludge and ensure its dissolution. Nevertheless, all anodes should be enclosed in closely-woven fabric bags to retain any sludge and prevent it being carried into the electrolyte. Depolarised or carbon-containing anodes are available in the cast or wrought form.

To an increasing extent, the use of such preformed anodes is being superseded by the introduction of titanium baskets containing nickel in the form of pellets or small, cut squares. Because of its insulating oxide film, titanium itself does not conduct current into the electrolyte, but at the local points of contact between the titanium and the nickel inside the basket, conduction is established and current flows through the nickel into the electrolyte. Active forms of nickel, containing about 0.02 per cent of sulphur introduced during the refining process, may be used to ensure smooth dissolution at a low potential, but normal primary carbonyl pellets or electrolytic nickel are also widely used. The complete titanium basket must be enclosed in suitable anode bags to retain sludge.

A wide variety of electrolytes have been proposed for nickel plating, but there are three basic solutions of major interest. The Watts-type bath is based on mixed sulphate and chloride solutions with an addition of boric acid. A typical composition is as follows:

nickel sulphate ($NiSO_4 . 6H_2O$)	240 g/l
nickel chloride ($NiCl_2 . 6H_2O$)	45 g/l
boric acid (H_3BO_3)	30 g/l

The solution can be operated over a wide range of temperatures and acidities, but at about 55°C and a pH of 3–5 it produces the softest, most ductile deposits with moderate levels of internal stress. Under these conditions the hardness and internal stress are related to the current density as shown in Fig. 55.

An all-chloride electrolyte is used which produces a finer-grained, smoother and less ductile deposit than the Watts bath. A typical composition is as follows:

nickel chloride	240 g/l
boric acid	24 g/l

This operates at 60°C and pH 2.0, with a current density of between 2.5 and 10 A/dm^2. It yields deposits with a hardness between 230 and 260 HV, a tensile strength of 680 MN/m^2 (69 kg/mm^2) and an elongation of 15 per cent.

For decorative coatings, particularly when a top coat of chromium is to be applied, a solution producing a bright deposit is a great advantage since no polishing is then necessary between the two plating operations. This can be obtained by making addi-

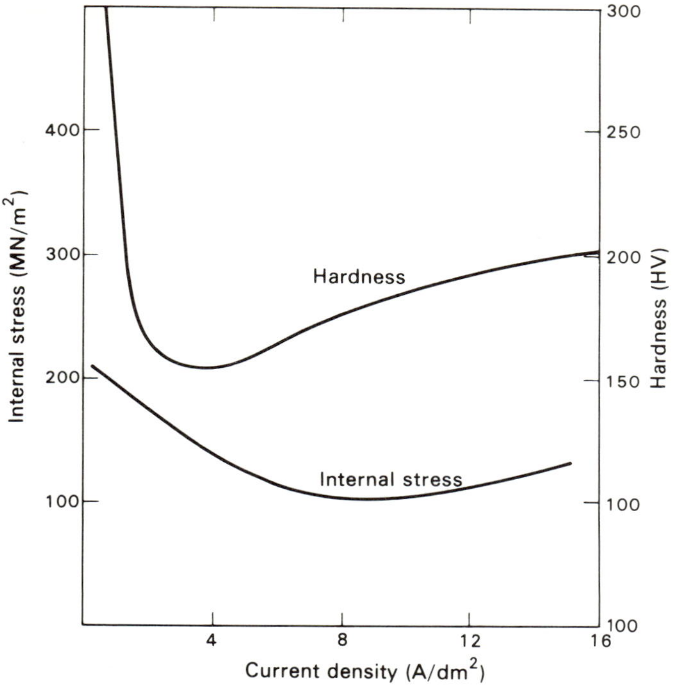

Fig. 55—Internal stress and hardness of nickel deposited from Watt's-type solutions operated at 55°C and pH 3.0.

tions of brightening or levelling agents to the electrolyte. The additions are usually organic compounds, sold as proprietary materials, and are effective in either Watts or all-chloride electrolytes. They usually involve a reduction in ductility of the deposit and an increase in the internal stress, so that at the same time an addition of a stress-control agent may also be desirable. The hardness of bright deposits will normally lie in the range 450–500 HV. The low ductility of bright nickel deposits makes them unsuitable for use on their own on thin sheet-metal parts subject to flexing in service under corrosive conditions. A double-layer plating system is therefore used for such parts, involving first a ductile layer of nickel followed by a bright nickel top layer. Corrosion penetrating through the cracked top layer then spreads laterally, rather than penetrating to the underlying base metal, because of the absence of cracks in the ductile nickel layer and also because this layer is more noble than the bright nickel layer.

While many articles, particularly those for domestic indoor use, are still plated only with nickel, the majority of plated articles are now finished with a top coat of chromium, about 0.25 μm thick. It must be recognised that the chromium coating itself contributes less than the nickel to the overall corrosion resistance, since, on account of its hardness and low ductility, it is invariably cracked to a greater or less extent. The underlying coating of nickel provides the major corrosion resistance and the level of this depends greatly on the thickness, as well as type, of deposit. Modifications to the conventional chromium deposit have been introduced, however, which improve its contribution to corrosion resistance by modifying the extent and nature of the crack pattern. One type provides a thicker layer, in the order of 0.75 μm, which is nominally crack-free; another provides what is termed "microcrack chromium", in which a very fine network of cracks (about 250 per cm) ensures that corrosion is widely dispersed instead of being concentrated at relatively few sites. Nevertheless, the nickel undercoat remains of critical importance. The controlling British Standard (BS 1224) defines the type of coating required to meet the following four conditions of service:

Grade 1. For articles exposed outdoors to severely corrosive conditions.
Grade 2. For articles used intermittently outdoors.
Grade 3. For articles used indoors where condensation may occur.
Grade 4. For articles used indoors in warm, dry atmospheres.

Depending upon the nature of the metal being coated (steel, copper-base alloys, zinc-base alloys or aluminium) and the nature of the nickel deposit (bright nickel, nickel requiring polishing, or double- or triple-layer nickel) the appropriate thickness of nickel to be used with different types of top coat is specified. Thicknesses of nickel range from 7.5–30 μm (3–12×10^{-4} ins.).

Heavy nickel deposits are frequently used to build up worn or damaged steel parts to recover the original dimensions[4], and for this process the two solutions mentioned earlier (p. 118) are commonly used. The rates of deposition, however, are relatively low, and in some cases continuous deposition over a period of several weeks may be required. Furthermore, the hardness of the deposits from these baths lies in the range 150–250 HV, but frequently harder, more wear-resisting coatings are necessary.

Modified electrolytes, usually containing ammonium salts, can give hardnesses up to 500 HV, a typical bath being as follows:

nickel sulphate ($NiSO_4 . 7H_2O$)	180 g/l
ammonium chloride (NH_4Cl)	25 g/l
boric acid (H_3BO_3)	30 g/l

This is operated at pH 5.6–5.9, at 43–60°C and with a current density of 2.5–5.0 A/dm^2. Depending on the exact conditions, the hardness can be from 350 to over 500 HV.

To obtain faster deposition rates than are available with the conventional Watts or chloride baths, advantage can be taken of the very high solubility of nickel sulphamate, $Ni(SO_3 . NH_2)_2$. A normal sulphamate electrolyte can yield deposits ranging in hardness from 150 to about 300 HV, and with appropriate addition agents can give very low, or indeed compressive rather than normal tensile, internal stresses[5]. Such a bath has the following typical composition:

nickel sulphamate ($Ni(SO_3 . NH_2)_2$)	300 g/l
nickel chloride ($NiCl_2 . 6H_2O$)	30 g/l
boric acid (H_3BO_3)	30 g/l

The bath can be operated over a wide range of conditions such as the following:

temperature	25–70°C
pH	3.0–5.0
current density	up to 15 A/dm^2 at 25°C up to 32 A/dm^2 at 70°C

Still higher deposition rates are obtainable by the "Ni-Speed" process which uses a similar electrolyte but with the nickel sulphamate content increased to 600 g/l. This is operated at temperatures between 60 and 70°C, and under suitable conditions of solution agitation, anode and cathode disposition, etc., current densities as high as 86 A/dm^2 can be used, corresponding to a nickel deposition rate of over 1 mm thickness per hour. The hardness of the deposits obtained from the Ni-Speed process depends on the current density as shown in Fig. 56. Further modification of the Ni-Speed process, by the inclusion of cobalt sulphamate in the electrolyte, enables nickel–cobalt alloy deposits to be produced with hardnesses up to 525 HV. With a cobalt ion concentration of 5–10 g/l, the deposits have cobalt contents in the range 30–50 per cent and hardnesses from 425 to 525 HV. Continuous and careful control of the composition of the electrolyte is necessary for the successful operation of the Ni-Speed process.

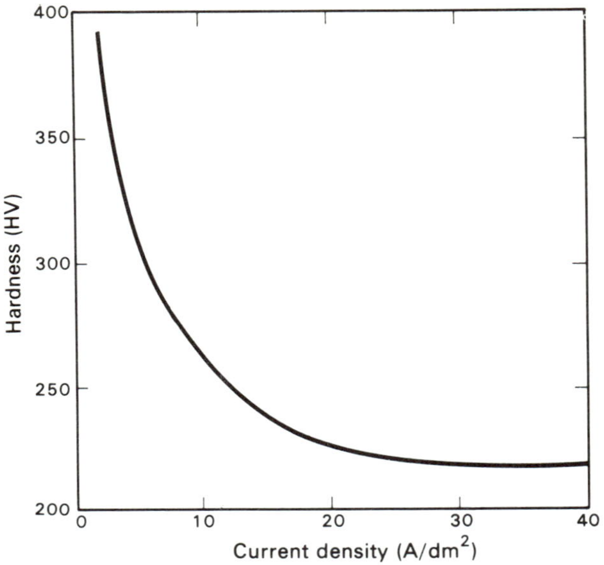

Fig. 56—Hardness of nickel deposited by the "Ni-Speed" process.

ELECTROFORMING

Electroforming is the production or reproduction of an article by electrodeposition onto a mandrel or former that is later removed. The process has been used for over a century for the production of low-relief sculptures, etc., mainly in copper, but in recent years nickel has been more widely used because of its superior mechanical properties. The products are finding applications in a wide field (*see* Chapter 7, p. 130). The particular advantages of the process are that intricate shapes can be reproduced with great fidelity and in fine detail, and the process can be adapted to large-scale production. The mandrels can be permanent or expendable. The former are usually metallic—stainless steel, copper, brass or nickel—and may be machined, engraved, or themselves electroformed from a permanent or expendable master. Plating with chromium may be desirable to aid separation of the electroform. Plastics are also used for permanent mandrels, notably for producing electroforms with surface textures, such as printing plates and embossing rolls. Expendable mandrels may be low-melting-point metals, soluble metals (for example, aluminium soluble in caustic soda), plastics or waxes. The metallic mandrels need to be carefully cleaned and treated to aid separation, while the non-metallic mandrels have to be made conductive by the application of a graphite coating or by the formation of a metallic film (for example, silver) by chemical reduction. Deposition of nickel to the required thickness may be made from the normal nickel-plating baths, provided the internal stress of the deposit is not so high as to cause distortion of the electroform. Modifications of the Watts bath are commonly used, and the sulphamate baths have the additional advantage of high deposition rates.

Electroless nickel deposition

Nickel may also be deposited by a non-electrolytic process based on catalytic reduction of nickel chloride. The reaction is catalysed by sodium hypophosphite added to the nickel chloride solution. The deposit consists of a nickel–phosphorus alloy containing 5–10 per cent phosphorus. It is hard, about 500 HV, the hardness increasing to 900–1000 HV after heating to about 400°C. Heating to higher temperatures reduces the hardness and improves ductility. The solution is operated at 85–100°C and deposits the alloy at a rate of about 15 μm/hour. The process has the advantage, compared with electrolytic deposition, of producing uniform thickness on all surfaces, and thus no problems of throwing power. The process operates with no pre-treatment on steel, aluminium, nickel or cobalt, but copper and brass are non-catalytic and deposition must be initiated by contact in the solution with one of the catalytic metals. Other metals may be coated similarly if first plated with copper. Non-metals may be coated by pre-treatment in palladium chloride solution.

References

1. *Machining Wiggin Nickel Alloys*, Henry Wiggin and Company Ltd., Publication No. 2463, Hereford, 1971.
2. *Welding, Brazing and Soldering Wiggin Nickel Alloys*, Henry Wiggin and Company Ltd., Publication No. 3367, Hereford, 1968.
3. Brenner, A., Zenter, V., and Jennings, C. W., "Physical Properties of Electrodeposited Metals: I, Nickel", *American Electroplaters Research Report* No. 20, *Plating*, vol. 39, pp. 865, 899, 933, 1952.
4. Oswald, J. W., *Heavy Electrodeposition of Nickel*, International Nickel Ltd., Publication No. 2471, 1962.
5. Kendrick, R. J., "High-speed Nickel Plating from Sulphamate Electrolytes", *Transactions of the Institute of Metal Finishing*, vol. 41, p. 59, 1964.

CHAPTER SEVEN

Applications of Nickel and the Nickel Alloys

The foregoing chapters have given an account of the wide variety of nickel-base alloys which are available to the engineer and have outlined their main properties and characteristics which are of value. This chapter reviews the uses to which these and other nickel-containing materials are applied, but it is of interest to note first the pattern of nickel consumption. Figures given by the International Nickel Company of Canada, now Inco Limited, in

TABLE 32

Consumption of nickel (1971).

Primary nickel by products (%)		*All nickel by end-use* (%)	
Stainless steels	41	Consumer and institutional products	17
Nickel plating	16	Machinery and transport	13
High-nickel alloys	12	Automotive	11
Constructional alloy steels	10	Petroleum	9
Iron and steel castings	9	Electronic	8
Copper and brass products	3	Chemical	8
Others	9	Process	7
		Aircraft	7
		Energy conversion	5
		Marine	3
		Building and construction	3
		Coinage	1
		Others	8
	100		100

their Annual Reports, indicate that consumption can be broken down into terms of products and end-use industries. These analyses are given in Table 32. The analysis in terms of products indicates that the steel and ferrous casting industries as a whole consume over 60 per cent of the total primary nickel used, and that stainless steel is by far the most important single product for nickel. When it is considered that the steel industry also uses considerable quantities of scrap material arising from other sources, such as from the nickel–chromium high-temperature alloys, to introduce nickel into their products, it is clear that an even higher proportion of total nickel consumption is accounted for by its use in steel. Nevertheless, it is not proposed to deal in any detail with these uses but to concentrate on the applications of the nickel-rich alloys which have been described earlier.

The analysis in terms of end-use indicates that the use of nickel-containing products, both steels and non-ferrous materials, is very widespread, and that the market does not depend to any major extent on a single end-product industry.

Nickel

The wrought and cast forms of nickel are of particular value in the chemical industry because of their resistance to corrosion in alkaline and other media. All the normal forms of the product are utilised, and for large-scale plant, nickel-clad steel is frequently used, the cladding being effected by hot rolling or by weld overlaying. Wrought nickel is used for evaporation vessels, heating coils, reaction chambers and other components used in the production of caustic soda, chlorinated hydrocarbons, fluorine, hydrofluoric acid and many less aggressive chemicals. Cast nickel, although used to a much smaller extent, finds some application in parts of such plant, particularly components for pumps and valves.

Many other applications based on the resistance of nickel to corrosion by the media involved are found in the chemical and process industries, and include plant for production and handling of petroleum; plastics; paper, pulp and textiles; salt evaporators and driers; food-processing equipment; dry-cleaning plant, etc.

The corrosion resistance of nickel is also one of the main reasons for its use as coinage. Other important characteristics in this application are the level of hardness when annealed and the work-hardening capability, which together make it both soft enough to take readily the impression from the dies, and, after striking, hard enough to have good wear resistance. Pure nickel coins were first introduced in Switzerland in 1881, and, although still used by a number of countries, have largely been superseded by coins in the 75/25 cupro-nickel alloy, which is now used for British "silver" coinage. Almost all coins are now only tokens, the metal value of the British 50p and 10p coins, for example, being only about 1.1p and 0.9p respectively.

The electrical and electronic industries provide many examples of the use of nickel in applications for which no other metal is equally suitable. In vacuum tubes of many types, ranging from small receiving valves to large high-power transmitting valves, television tubes and other electronic devices, nickel is used for anodes, grids, cathodes and support wires. Special grades of the metal with minor contents of other metals are made to meet the particular requirements of the individual parts, in terms of strength, ductility, resistance to sulphur embrittlement and electronic emission characteristics.

Nickel alloys with additions of manganese and silicon up to about 4 per cent and 1 per cent respectively, and with small additions of a reactive element such as zirconium to control the grain size, are used for the electrodes of internal-combustion engine sparking plugs. The alloy must be resistant to the corrosive action of lead oxide and bromide deposited from the anti-knock agent added to the fuel, and to spark erosion; it must be of adequate strength and yet sufficiently ductile to permit gap setting by bending.

The magnetostrictive properties of nickel are used in a number of practical devices, the most important of which are transducers for the generation of ultrasonic energy. These consist basically of a core of the magnetostrictive material designed so that its natural frequency of longitudinal vibration is the required frequency of ultrasound, placed inside a coil to which alternating current of the appropriate frequency is fed. The core is normally made from laminations 0.1–0.25 mm thick to minimise eddy-current losses, and may be oxidised under controlled conditions to produce a thin insulating oxide film, thus avoiding the need to use insulating varnishes between the laminations. Ultrasonic generators are used in underwater echo-sounding devices in which a pulse of ultrasound is emitted, and the time delay before the receipt of its echo, by the same or another transducer, is measured. Depth sounding for hydrographic survey and navigational purposes, fish location and naval underwater vessel location, are well-established applications of echo-sounding equipment. Ultrasonics generated by magnetostrictive transducers are used in industrial cleaning operations, sometimes using very high-power sources. A growing application is in the ultrasonic soldering iron, used with aluminium and other metals carrying a strong, coherent oxide film. The vibrations mechanically help to detach the film. The common use of ultrasonics for flaw detection in metals normally operates at very high frequencies for which piezoelectric crystal transducers are more suitable, but magnetostrictive devices are preferred for medium or lower frequencies, or when large power outputs are necessary.

The magnetostrictive behaviour of nickel is applied in "delay lines" used in computers. A pulse of mechanical energy is introduced magnetostrictively at one end of a wire and is reconverted to an electrical impulse on arrival at the other end, by the reverse effect. The time delay depends on the length of the wire and may be of the order of 10^{-5}–10^{-3} seconds.

Nickel powder

As with other forms of primary nickel, a major outlet for nickel powder is as an alloying constituent in steel parts—sintered-steel components may contain up to 5 per cent nickel. The advantages conferred by the nickel are, as for wrought alloy steels, improved hardenability, strength and toughness. Nickel powder is, however, used without alloying in a number of applications which exploit the corrosion resistance of the element, and two of these are concerned with structures with controlled porosity.

The first is as a porous electrode in storage batteries and fuel cells using alkaline electrolytes. Nickel-alkaline cells all use nickel hydroxide as the active material of the positive plate, the electrolyte being an aqueous solution of potassium hydroxide. The negative plate may be cadmium or iron. There are several different basic designs of the positive plate, some using nickel-plated steel or pure nickel foil as the structural material, with the active hydroxide held in pockets or packed into tubes. The sintered type of plate is made by sintering nickel powder around a reinforcing grid of nickel or nickel-plated steel, to give about 80 per cent porosity. The active mass is introduced into the voids by a chemical impregnation process. The advantage of the sintered plate is that the active mass is in intimate contact with the nickel skeleton structure over a high surface area, so that the cell has low internal resistance, can maintain high rates of discharge, and can be recharged rapidly. Batteries containing this type of plate are used in aircraft-engine starting, transportable power sources, and small sealed cells operating a wide variety of hand tools, instruments and devices.

Fuel-cell electrodes also consist of porous plates, but no active mass is incorporated in the pores. The fuel, usually hydrogen or other clean gas, is fed to one side of the plate, on the other side of which is the electrolyte. The electrochemical reaction takes place at the meniscus within the pores of the plate and the electricity released is collected by the metallic electrode. The electrode may have graded porosity through its thickness to ensure that the meniscus is retained within the electrode. Figure 57 shows a photomicrograph of a triple-layer electrode.

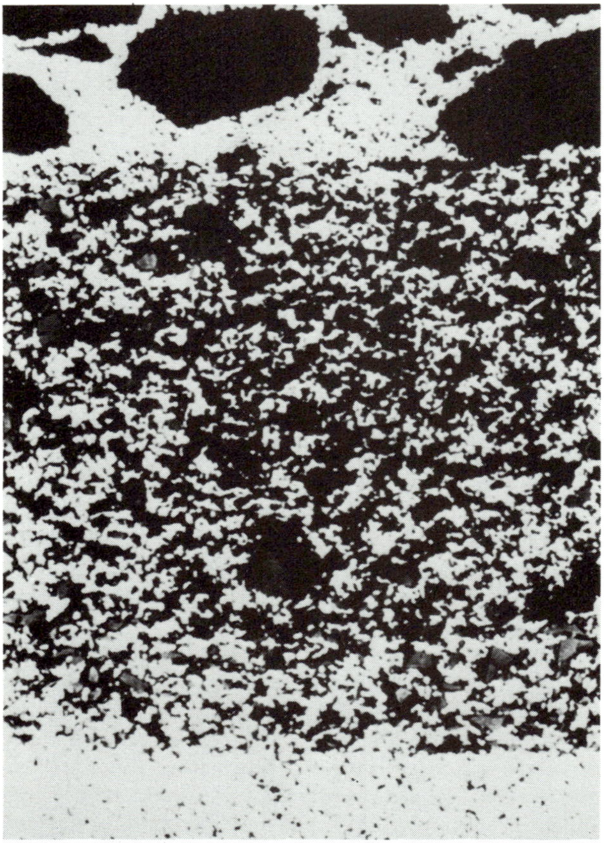

[*Varta A.G.*

Fig. 57—Three-stage graded-porosity fuel-cell electrode ($\times 200$).

The second major application of porous sintered nickel is as a filter or diffusion membrane. In particular, components in the form of tubes have been used in nuclear plant as diffusion membranes for the separation of uranium isotopes combined as the hexafluorides.

Fully-dense nickel, made by continuous rolling of nickel powder, has been used for coinage. A disadvantage of this method of production is that the scrap, after punching the coin blanks from the rolled strip, cannot readily be recirculated in the same production system; an alternative use for the scrap metal must be found.

Nickel plating

Electrodeposited nickel is widely used on a variety of base metals, to provide protection against corrosion, a bright decorative finish (in association with chromium), and heavy coatings for purely engineering reasons.

Many small articles for domestic or general indoor use are nickel plated; these are so numerous that it is sufficient to say that almost any metallic article, for which special mechanical or physical properties are not required, can be more economically produced from nickel-plated mild steel than from other metals—pins, paper clips, scissors, buckles, fasteners, keys—the list could be extended almost indefinitely. The thickness of nickel required for moderate service of this nature is small, and generally need not exceed 0.25–0.5 μm. In specifically corrosive conditions, where the aim of the plating is not merely to prevent the surface from tarnishing or rusting in the atmosphere, but to protect the underlying metal from more aggressive media, much thicker, pore-free coatings are required. For example, in food preparation and processing plant, the paper and pulp industry, and the chemical and pharmaceutical industries, nickel plating is widely used; thicknesses up to 25 μm or more may be required.

By far the greater part of purely decorative plated coatings nowadays consists of bright chromium applied over an undercoating of nickel, or copper plus nickel. The undercoats provide the essential corrosion protection of the basis metal, while the chromium layer prevents surface tarnishing of the nickel. The type and thicknesses of the coatings required for varying service conditions are given in appropriate specifications (for example, BS 1224, 1965). The applications of nickel–chromium plating are so ubiquitous that examples are unnecessary.

Heavy nickel plating, apart from providing corrosion protection, may be used to salvage worn, corroded or incorrectly machined parts, or to provide a surface coating more suitable than the bare steel for the required service, which may involve frictional or erosion characteristics. For example, heavy nickel plating has been used on one member of close-fitting joints to prevent fretting.

Applications of nickel electroforming are many, and it is a rapidly increasing technique for the reproduction of articles of complex form. Perhaps the best established is the use of electroformed stampers for the production of gramophone records. Dies for repetitive plastic-injection moulding may be electroformed from master patterns. Precision articles such as tubing, bellows, nozzles, radar wave-guides, etc., are reproduced with accuracy and consistency by electroforming onto either permanent or expendable mandrels, depending on the shape. Screens, grids and filters of many types may be produced economically and consistently by electroforming onto a suitable mandrel, and in some cases production of continuous strip may be effected from a rotating mandrel. Normally, a stainless-steel mandrel is used and the pattern of holes is formed by applying a photo-sensitive resist and optically projecting the required pattern. After developing the image and etching it into the stainless-steel surface, the etched recesses are filled with a non-conducting material so that during subsequent electrodeposition of nickel these parts remain bare. Screens with very fine mesh patterns have been made, and are used for filters in food processing, the oil and paper industries, for fabric printing screens and for colour television screens.

NICKEL–COPPER ALLOYS

The nickel–copper alloy Monel 400 is more corrosion resistant than nickel under reducing conditions and more resistant than copper under oxidising conditions. These characteristics (together with its higher strength, very adequate ductility and ease of cold working) make it generally more attractive than either of the pure metals for a wide variety of applications, nearly all of which exploit its corrosion resistance.

In fresh water its uses range from small delicate parts in water meters to weld-deposited coatings on extremely large gate valves controlling water flow to power turbines; it is also used widely in the form of tubing for feed-water heaters in power stations and marine boilers. In salt and brackish waters its uses are also widespread, its resistance to cavitation and erosion in high-velocity water streams often being important. Heat-exchanger tubes, pump parts, marine propellers and propeller-shafts are typical applications, although for the latter component the age-hardening version of the alloy may be preferred on account of its improved strength. Nickel–copper alloys are not so satisfactory in stagnant sea-water since pitting corrosion may arise where marine organisms can grow and produce oxygen-concentration cells. Its resistance to brine makes it suitable for sea-water evaporators, either in desalination plants or in salt recovery plants. In most heat-exchanger units of this type the alloy is used in the form of seamless solid-drawn tube.

In the chemical and process industries the alloy is used for vessels, pipe-lines, filters and valves in many different media. Its availability in almost all forms, and the readiness with which it can be formed and welded, make it an important candidate for many applications. Thus in the production of ammonium sulphate for fertilisers the salt is dried in baskets made of the alloy; the processing of animal and vegetable oils and fats, often involving refining treatments with sulphuric acid, may be carried out in plant using nickel–copper alloy for many components—tanks, pipe-lines, heating coils, filter screens, etc. In food-processing equipment generally, particularly when salt solutions are involved, the alloy is commonly used. Filters are made in several forms—woven-wire cloths, space-wound wires or perforated sheet or foil are all used in such industries as paper making, salt production, sugar refining, oil refining and sewage treatment.

In the metallurgical and engineering industries, the alloy is used in the pickling of iron and steel. The acids used are commonly relatively dilute mixtures of sulphuric and hydrochloric acids at normal or only moderately elevated temperatures. Nickel–copper alloys are suitable for crates, hooks, baskets, etc. Figure 58 shows typical pickling equipment. Similar equipment is used in the glass industry for handling glassware during pickling in hydrofluoric acid. Laundry and dry-cleaning equipment is another field in which the corrosion resistance of nickel–copper alloy is exploited, the alloys being adequately resistant to attack by chlorinated solvents and their breakdown products.

Nickel–copper alloys are used for a variety of small articles in the electrical industry on account of their satisfactory mechanical properties associated with corrosion resistance and heat resistance. Terminals, washers, contacts, connecting leads, etc., are all used where corrosion or oxidation might arise if a less resistant material were used. Monel alloy 400 is also used in large quantities for pop rivets used as fasteners in many articles in the electrical

[*Henry Wiggin and Co. Ltd.*

Fig. 58—Hangers and racks in Monel alloy 400 in continuous pickling plant.

and light-engineering industries. Similarly, it is used in electronics for such articles as transistor capsules, and also as wave-guides, in which ease of formability and capability of taking a high surface finish are important characteristics.

COPPER–NICKEL ALLOYS

These alloys generally find applications in the same fields as the higher-nickel alloys referred to above, but generally under conditions of service which are rather less severe, and hence do not justify the higher cost of the nickel-rich alloy. It must be remembered that the market price of primary nickel is usually between two and three times that of copper.

Cupro-nickels with either 10 or 30 per cent nickel are particularly important in the field of heat-exchangers involving sea-water—either as feed-water heaters or condensers. The alloys are mainly used in the form of solid-drawn tubing, but sheet and castings are also employed in such components as water boxes, trunking, valves, etc. In addition to their long-established use in marine power-plant, the alloys are now finding increasing use, under basically similar conditions, in desalination plant for the production of fresh water from sea-water. Apart from these applications involving sea-water, heat-exchangers of a similar character are important in the chemical industry and, depending on the reagents to be handled, cupro-nickels may be the most suitable construction alloy.

NICKEL-SILVER

Many of the applications of nickel-silvers are still based on their attractive white colour and associated with their ease of forming and fabrication, which were the main features leading to the original use of the alloy for domestic table-ware, hollow-ware and flat-ware. Much of the early production of nickel-silver articles of this sort was silver plated, and E.P.N.S.-ware (electro-plated nickel-silver) effectively supplanted roll-bonding of silver on a copper base (Sheffield Plate) over a century ago. The use of these alloys in this field still continues, in spite of the increasing competition from stainless steel. Both cast and wrought forms of the alloys are also used in architectural fields, mainly as high-quality decorative fittings such as hand rails, door furniture, etc. An increasing quantity is being used, however, for applications based purely on the technical rather than aesthetic merits of the materials. These are mainly in the electrical industry where the range of mechanical properties obtainable by cold working (coupled with the corrosion resistance and the good sliding-contact behaviour) are particularly meritorious. Perhaps the most important application in this industry are the flat springs in telephone relays. Sliding electrical contacts in switches, plug and socket connections, etc., are frequently made in nickel-silver, and the corrosion resistance is such that, provided sufficient sliding action is present to break through the light tarnish film, there is no need for protection by plating or other means. Other common applications which depend on the ease of forming and machining of the alloys, and on their good low-friction characteristics, are cylinder-lock keys, zip-fasteners and hinges of spectacle frames. For all these applications, a lead-containing version of the alloys is usually preferred.

NICKEL–CHROMIUM AND NICKEL–IRON–CHROMIUM ALLOYS

The binary 80/20 nickel–chromium alloy is perhaps the most generally satisfactory alloy for resistance to high-temperature oxidation, and in the form of wire or tape is used for the element of many electrically-heated devices. The alloy is serviceable at temperatures as high as 1150°C, but few applications other than certain industrial furnaces operate above about 1000°C. Indeed, a great many domestic, electrical heating appliances operate at much lower temperatures, so that more economical alloys such as those in the nickel–iron–chromium system, typically 60 per cent nickel, 20 per cent iron and 20 per cent chromium, may be satisfactory. The only significant competition to nickel–chromium or nickel–iron–chromium alloys for the higher-temperature elements comes from alloys of the iron–chromium–aluminium system, which under suitable conditions may operate at slightly higher temperatures than the best nickel–chromium alloys. However, they suffer from two disadvantages: firstly, their high-temperature strength is less than that of the nickel–chromium alloys and hence the heated wires may distort under their own weight, causing local overheating and early failure unless very fully supported; and secondly, they embrittle after heating to high temperature, so that they cannot later be cold-formed or adjusted in position.

The choice of element material will usually be made quite readily in the light of the operating temperature of the element. Such articles as radiant fires, toasters, etc., in which the element normally reaches perhaps 850–900°C, should use the binary nickel–chromium alloy to allow for occasional or localised overheating without abnormal shortening of the life. For lower-temperature elements for articles such as immersion heaters, fan heaters, hair driers, etc., the nickel–iron–chromium alloy may be quite satisfactory, its maximum permissible service temperature being normally regarded as 950°C, compared with 1150°C for the binary alloy. For the sheath material of enclosed elements such as are used for cooker hot-plates, a nickel–iron–chromium alloy is usually chosen. Inconel alloy 600 and Incoloy alloy 800 are both used, and either alloy gives good service life; the choice depends on the particular manufacturing technique to be adopted (usually involving automatic welding) as well as on economic factors.

For industrial furnace elements, the binary nickel–chromium alloy is the usual choice (in this case without the addition of reactive constituents to give resistance to scale shedding) and these are normally made from strip up to about 20–30 mm wide and 5 mm thick. Elements of this type in a vitreous enamelling furnace are shown in Fig. 59. When carburising conditions are likely to arise, particularly if they are intermittent and perhaps alternating with oxidising or reducing conditions, there is risk of

[*Henry Wiggin and Co. Ltd.*

Fig. 59—Industrial enamelling furnace with nickel–chromium heating elements.

the elements suffering green-rot attack (*see* Chapter 3, p. 50 and Fig. 44). A nickel–iron–chromium alloy should then be chosen with about equal proportions of nickel and iron and with a high silicon content (for example, Incoloy alloy DS). Although the maximum service temperature of the material is reduced to about 1050°C, this is usually adequate. Alloys of the nickel–iron–chromium system of both the Inconel and Incoloy groups are generally chosen for industrial furnace components of all types apart from the actual electrical elements. Thus retorts and muffles, nitriding boxes, heat-treatment trays, support hooks, etc., will generally be made in Inconel alloy 600 if the temperature is expected to exceed 900–1000°C or the longest life is required. Otherwise, Incoloy alloy 800 would be used, or for carburisation resistance the high-silicon version referred to above. All these alloys are available in all the common wrought forms, and are readily weldable so that no constructional difficulties arise.

For resistance to severe hot corrosion arising from the presence of fuel ashes, binary nickel–chromium alloys with chromium contents in the range 30–50 per cent are superior to the lower-chromium alloys. This type of corrosion occurs in boilers and furnaces burning residual fuel oils or pulverised coal, and was described in Chapter 3, pp. 45–47. The corrosion is suffered principally by the tube support brackets and hooks, which are not cooled as are the tubes themselves.

CREEP-RESISTING ALLOYS

Nickel-base creep-resisting alloys were developed from the binary nickel–chromium alloys to meet the needs of the aircraft gas-turbine engine in the early days of its development. Indeed, the success of the engine depended to a large extent on the provision of alloys with properties much superior to those of the materials previously available. The critical components are the turbine rotor blades which operate at temperatures in the range 700–1000°C under high stresses compounded of centrifugal stress due to the high rotational speed, gas bending stresses and vibrational stresses. Rotor blades are produced by a variety of methods, some being machined from forgings or wrought bar, and others being cast to size by precision methods involving little subsequent machining. Internal air-cooling passages may be formed in either cast or wrought blades.

The stator blades or nozzle guide vanes operate at higher temperatures than the corresponding rotor blades but are much less highly stressed—their main requirement is corrosion resistance and resistance to thermal shock and fatigue arising from change in operating conditions. These components are usually precision cast and may be subjected to surface coating procedures to improve resistance to hot corrosion, particularly for service in marine conditions.

The combustion systems of aircraft engines are fabricated from sheet material, and for this the solid-solution alloys such as Nimonic alloy 75 or Hastelloy X are largely used. The main requirements are reasonable hot strength and resistance to thermal shock, combined with ease of working and weldability without the need for subsequent heat treatment.

With the increasing operating temperatures of aircraft gas-turbines, the use of nickel-base superalloys is being extended to other components which were previously made in heat-resisting or alloy steels. In particular, rotor discs, shafts, rings, casings and jet pipes may all be made of nickel-base alloys, and frequently alloys are developed with characteristics specially suited to the individual components. Thus Nimonic alloy 901 is mainly used for rotor discs.

Nickel-base superalloys are widely used for the corresponding

[*Henry Wiggin and Co. Ltd.*

Fig. 60—Single-piece gas-turbine rotor in Nimonic alloy 90, diameter approximately 300 mm.

critical components of other gas turbines. For transport purposes, both marine and land, weight considerations are still important so that in general the design follows the lead of the aircraft industry. For turbines of relatively small power, such as are used for transportable power sources and are under development for road-transport purposes, the tendency is to use an integral rotor, the disc and blades being machined from a single forging or cast in one piece. Such a rotor is shown in Fig. 60. For larger stationary turbines for power generation, the alloys are used for rotor blades, usually as fully-machined oversize forgings, and for various sheet components such as combustion chambers, ducts, etc.

The success of nickel-base creep-resisting alloys in the gas-turbine field has led to their use in many other areas where their high-temperature strength is advantageous. Thus as exhaust valves in internal-combustion engines, their strength ensures freedom from stretching at the neck, and their corrosion resistance almost eliminates burning of the seat. In the steam-power field, materials with high strength at operating temperature are required for casing bolts, to minimise stress relaxation and consequent leakage at the casing joints. Such bolts or studs may be 100 mm or more in diameter and over one metre long, and operate at temperatures up to about 600°C. Nimonic 80A bolts have been used with considerable success. In the metal-working field there are clear advantages to be obtained from the use of high-strength heat-resisting alloys: die-casting moulds, extrusion dies, forging-press anvils and tools, hot-shear blades, etc., all have longer lives when made from appropriate nickel-base superalloys instead of the more conventional special steels, Although the mechanical stresses may not be very high on such tools, contact with hot metal generates very high thermal stresses. The improved life obtained with the nickel-base alloys is largely due to their high thermal shock-resistance.

NICKEL–IRON ALLOYS

Soft magnetic alloys

The applications of high-permeability nickel–iron alloys cover a wide range of devices in the electronic and telecommunications field. One of the earliest and most important of these was the use of Permalloy for the magnetic loading of telephone cables to increase the inductance. This was effected either by a continuous wrapping of high-permeability alloy tape outside the copper conductor, or by the insertion at intervals of loading coils wound around high-permeability cores.

Cores made from sheet laminations are also used for transformers and inductances used in audio-frequency work—much smaller cores in the nickel–iron alloy give an equivalent or better performance than larger and heavier cores of silicon–iron. For some telecommunication purposes, powder cores are used instead of laminations, giving reduced hysteresis and eddy-current losses, even though the effective permeability is not so high. Nickel–iron powders are used as well as pure iron powders.

Magnetic screens for electronic equipment are usefully made of high-permeability material since the shielding effect is directly proportional to the permeability.

Armatures and pole pieces of high-frequency generators, moving-iron instruments, pick-ups and magnetic recorders, etc., may all be made with high-permeability materials, usually in the form of laminations stamped from sheet, or as wound tape.

Permanent magnets

Permanent magnets are familiar to all and have many well-established uses as well as a multiplicity of unusual ones. Loudspeakers provide one of the main applications, where they have now completely displaced electromagnetic speakers in radio and television sets. The focusing of television tubes is also effected by permanent magnets. Telephones, microphones and electrical measuring instruments provide a very large market for magnets. In the instrument field, the stability of modern, high-energy magnets has been a significant factors in improving the accuracy and reliability of instruments, at the same time reducing their weight and bulk. In addition to providing the main magnetic field in all instruments, they find an application as eddy-current brake magnets in watt-hour meters.

Permanent magnet motors and generators of a wide range of sizes are made for use on motor-cycles, automobiles, toys, servo-mechanisms, etc.; large d.c. motors may also employ permanent magnets.

The mechanical action of permanent magnets is used in many applications. Switches, relays, drives and catches of various types are designed to depend on magnetic attraction or repulsion. Magnetic chucks energised by permanent magnets can be as powerful as electromagnetic chucks and have the advantages of requiring no current leads and generating no heat. They can be switched on and off by a simple lever. Figure 61 illustrates such a chuck.

By far the majority of permanent magnets are made from nickel-containing alloys of the types described in Chapter 3, p. 69.

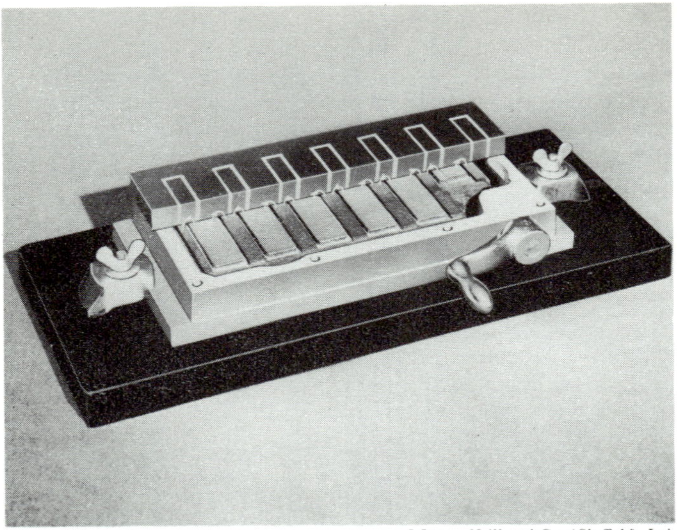

[*James Neill and Co. (Sheffield) Ltd.*

Fig. 61—Cut-away model of permanent magnetic chuck.

Controlled expansion alloys

The controlled expansion alloys also find a very wide variety of applications in the electrical and engineering industries. The low expansion of Invar at temperatures near to ambient is of great importance in metrology and is used for standard scales, measuring tapes, chronometer pendulums, etc. The differential expansion of materials which need to be clamped together can be compensated for by introducing a spacing piece or washer in a low-expansion nickel–iron alloy—such a system has been used, for example, between steel and aluminium members in a cryogenic liquid pipe-line.

The alloys selected to have expansion coefficients matching that of glass are used for many devices involving glass-to-metal joints. Lamp bulbs, fluorescent tubes, radio valves, television tubes and other vacuum equipment all involve sealed joints of this type. Reed switches consist of two or more wires of nickel–iron alloy sealed into a glass envelope and operated by an external solenoid or permanent magnet. Both the magnetic and expansion characteristics of the alloy are important in this application. Low-expansion alloys form one constituent of bi-metallic strip used for thermostatic purposes; the other member is commonly an austenitic steel.

The two materials are normally roll-bonded to form the strip. With changing temperature the strip curves, due to differential expansion, and provides the measuring or activating movement of the device. It is used in thermostats for space heaters, cookers, domestic irons, etc., and for thermometers, starter switches for fluorescent tubes, overload circuit breakers, etc. A typical application is the dial thermometer shown in Fig. 62.

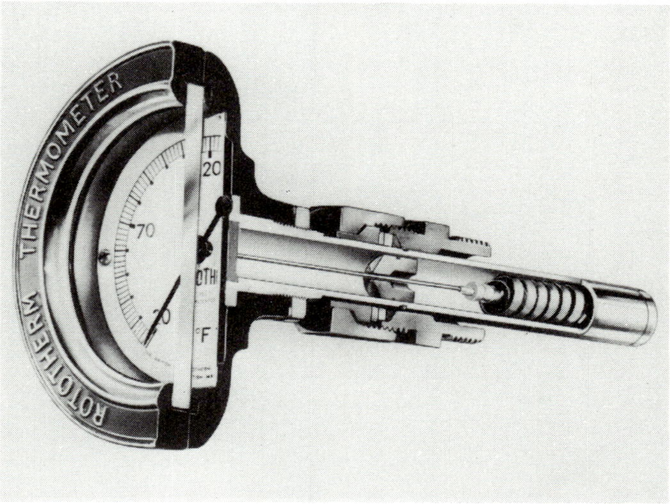

[*The British Rototherm Co. Ltd.*

Fig. 62—Dial thermometer with nickel-containing bi-metallic helical element.

CONSTANT MODULUS ALLOYS

These alloys are used principally for the hair-springs of clocks and watches, thus avoiding the need for elaborate temperature compensating devices. They are also used in precision spring-balances and weighing machines. More sophisticated instruments which rely on the constant frequency of a vibrating reed, made of an appropriate alloy, may also use the magnetic or magnetostrictive characteristics for driving purposes.

APPENDIX ONE

SI Units in the Industrial Metals Series

Quantity	*SI unit*	*Recommended multiples*	*Other related units (or names)*
Length	m (metre)	km m mm nm	Å = 10^{-10} m
Area	m^2	mm^2	hectare (ha) = 10^4 m^2
Volume	m^3	mm^3	litre (l) = 10^{-3} m^3
Time	s (second)	s ms ns	minute (min) hour (h) day (d)
Mass	kg	g mg μg	tonne (t) = 10^3 kg metric carat = 2×10^{-4} kg
Density	$\frac{kg}{m^3}$	$\frac{g}{cm^3}$	
Force	N (newton)	MN kN	dyne = 10^{-5} N
Impact strength	$\frac{J}{m^2}$	$\frac{kJ}{m^2}$	$\frac{J}{cm^2}$
Stress and pressure	$\frac{N}{m^2}$	$\frac{MN}{m^2}$ $\frac{N}{mm^2}$	hectobar (hbar) = 10^7 N/m^2 pascal = N/m^2

Quantity	*SI unit*	*Recommended multiples*	*Other related units (or names)*
Viscosity	$\frac{Ns}{m^2}$		centipoise (cP) $=10^{-3}$ Ns/m^2
Surface tension	$\frac{N}{m}$	$\frac{mN}{m}$	
Energy and work	J (joule)	GJ MJ kJ	kilowatt hour (kWh) $=3.6\times10^6$ J $=3.6$ MJ electronvolt (eV) $=1.602\ 19\times10^{-19}$ J
Power	W (watt)	GW MW kW	
Temperature	K		°C (Celsius)
Linear expansion coefficient	$\frac{1}{K}$		$\frac{1}{°C}$
Heat quantity	J	MJ kJ mJ	calorie $=4.1868$ J
Thermal conductivity	$\frac{W}{mK}$		$\frac{W}{m°C}$
Specific heat capacity	$\frac{J}{kgK}$	$\frac{kJ}{kgK}$	$\frac{J}{kg°C}$
Latent heat	$\frac{J}{kg}$	$\frac{kJ}{kg}$	
Electric current	A (ampere)	mA	
Electric potential	V (volt)	MV kV mV	
Current density	$\frac{A}{m^2}$	$\frac{A}{cm^2}$ $\frac{A}{mm^2}$	
Resistance	Ω (ohm)	mΩ $\mu\Omega$	
Resistivity	Ωm	$\mu\Omega$m	$\mu\Omega$cm $=10^{-8}\ \Omega$m

APPENDIX TWO

Common Conversion Factors

1 yard	=0.9144 m
1 foot	=0.3048 m
1 inch	=25.4 mm
1 mil	=0.001 in=25.4μm
1 yd^2	=0.836 127 m^2
1 ft^2	=0.092 903 m^2
1 in^2	=6.4516 cm^2
1 yd^3	=0.764 555 m^3
1 ft^3	=28.3168 dm^3
1 in^3	=16.3871 cm^3
1 gal (Imp.)	=4.5461 litre or =4.546 09 dm^3
1 in^4	=41.6231 cm^4 (Moment of Section)
1 troy oz	=31.1035 g
1 dwt (pennyweight)	=1.555 17 g
1 oz (av)	=28.3495 g
1 lb	=0.453 592 kg
1 cwt	=50.802 kg
1 ton	=1016.05 kg
1 tonne (metric)	=1000 kg
1 lb/in^3	=27.6799 g/cm^3
1 lbf	=4.448 22 N
1 tonf	=9964.02 N
1 tonf/in^2	=15.4443 MN/m^2 or =15.4443 N/mm^2
1 assay ton	=32.6667 g

Index

236/299.—